# Fundamentals of
# Limnology

# The Authors

**Jayashree Datta Munshi** was Professor of Botany in the Department of Botany, T.M. Bhagalpur University. She has done Extensive Research Work on different aspects of Plants, diarh lands, forests and limnology of ponds, reservoir lakes, thermal springs, wetlands river. She was principal investigator in the several U.G.C., CSIR and sponsored research projects.

**Jyotiswarup Datta Munshi** born 8 February, 1930, Ph.D Banaras Hindu University (1959), Professor and Head, Post-Graduate Department of Zoology, Bhagalpur University 1970-1992. Formerly, Lecturer in Zoology, Patna University (1952-62), Reader in Zoology, Banaras Hindu University (1962-70), Professor Datta Munshi continued to be associated with Bhagalpur University as CSIR Emeritus Scientist (1992-95) and Senior Scientist, INSA (1996-2000). Prof. Datta Munshi has been the recipient of British Council Visiting Lectureship (1971,1977), INSA-Royal Society Exchange Programme Visiting Professorship (1985) at Bristol University, U.K., Max-Planck Institute Research Fellowship, Germany (1985), Visiting Scientist under INSA/DFG Germany Exchange Programme (1997), INSA/Polish exchange programme (1991), Rockefeller Foundation Programme (1989), Nuffield Foundation and Smithsonian Foundation Visiting Scientist (1984, 87, 89,90,94) at University of Notre Dame, U.S.A.

Prof. Datta Munshi's outstanding achievements, have earned him several honours such as Patna University Gold Medal (1952), S.L. Hora Gold Medal (1980), INSA, Chandrakala Hora Memorial Medal (1992), Golden Jubilee Medal of Zoological Society, Calcutta (1997), 20th Century Gold Medal of Zoological Society of India, (1998). Prof. Datta Munshi is a Fellow of Indian National Science Academy (FNA), National Academy of Sciences (FNASc) and Zoological Society of India (FZA).

**Societies**: Fellow of National Academy of Sciences, India; Fellow of the Indian National Science Academy (1980- ); member, Ichthyological Society of Japan and Society for Experimental Biology, London; Life Member, Indian Science Congress Association.

**Award and Honours:** Patna University Gold Medal (1952), S.L. Hora Gold Medal (1980), Indian Society of Ichthyologists, Member, INSA Council; Bihar Gaurav Puraskar 2007, (2007).

Professor Datta Munshi has contributed significantly to the understanding of the structure, function and evolution of the air-breathing organs of teleostean fishes. His morphometric analyses of the bimodal respiratory system have indicated how changes in their relative development can be correlated with changes in the life-pattern of the fish, and seasonal characteristics of the environment.

**Address:** P-124/A, usha Park, Garia, Kolkata – 700 084.

# Fundamentals of
# Limnology

Jayashree Datta Munshi
and
Jyotiswarup Datta Munshi

2015
## Daya Publishing House®
*A Division of*
## Astral International Pvt. Ltd.
New Delhi – 110 002

**Cataloging in Publication Data–DK**
    Courtesy: D.K. Agencies (P) Ltd. <docinfo@dkagencies.com>
    **Datta Munshi, Jayashree**, 1940- **author**.
Fundamentals of limnology / Jayashree Datta Munshi and Jyotiswarup Datta Munshi.
        pages    cm
        Includes bibliographical references (pages      ) and indexes.
        ISBN 978-93-5130-690-0 (International Edition)

        1. Limnology–India. 2. Fisheries–India. 3. Aquatic ecology–India. 4. Freshwater ecology–India.    I. Datta Munshi, J. S., author. II. Title.

    DDC 551.4820954      23

*Published by*            :  **Daya Publishing House®**
                             *A Division of*
                             **Astral International Pvt. Ltd.**
                             – ISO 9001:2008 Certified Company –
                             4760-61/23, Ansari Road, Darya Ganj
                             New Delhi-110 002
                             Ph. 011-43549197, 23278134
                             E-mail: info@astralint.com
                             Website: www.astralint.com

*Laser Typesetting*       :  Classic Computer Services, **Delhi - 110 035**

*Printed at*              :  Thomson Press India Limited


**PRINTED IN INDIA**

The book is in the memory of
*My Parents*

**Sachindra Nath Datta**
**&**
**Santilata Datta**

# Acknowledgement

I am indebted to a number of International Collaborators and National Scientists A.B. Mishra, Professor and Head of Department of Zoology, Banaras Hindu University and Research Scholars for carrying out research work with me on different aspects of structure, function, evolution, eco-physiology and phylogeny of fishes of India.

## International Collaborators

1) Professor George Morgan Hughes, Research Unit for Comparative Animal Respiration, Bristol University, England, U.K.

2) Professor Ewald R. Weibel, Anatomical Institute, University of Berne, Switzerland.

3) Professor Kenneth R. Olson, Indiana University School of Medicine, South Bend Centre, University of Notre Dame, Notre Dame, Indiana – 46556, U.S.A.

4) Professor Hiran M Dutta, Kent State University, Ohio, U.S.A

5) Professor Peter Gehr, Anatomical Institute, University of Bern, Switzerland.

6) Professor Johannes Piiper, Max-Planck-Instut fur experimentelle Medizin, Abteilung Physiologie, Gottingen University, Germany.

7) Professor Pierre Dejours, Laboratoire d'etude des régulations physiologiques (associéa I' université Louis Pasteur) Centre national de la recherché scientifique, France.

8) Professor Giacomo Zaccone, Department of Animal Biology and Marine Ecology, Faculty of Science, University of Messina I-91866, Messina, Italy

## Research Fellows

Bali Ram Singh, Bans Narain Singh, Satyendra Prasan Singh, Ajay Kumar Mittal, Subhas Chandra Dube, Ram Kumar Singh, Jagdish Ojha, Bhupendra Narayan Pandey, Syed Aftab Khalid Nasar, Mahadeo Prasad Saha, Narendra Deo Prasad Sinha, Narendra Mishra, Abdul Hakim, Asha Lata Sinha, Ajay Kumar Patra, Niva Biswas, Devendra Prasad Choudhary, Dayanand Roy, Surendra Prasad Roy, M.A.O. Johar, Prem Kumar Verma, Shobha Sah, Arun Kumar Laal, Upendra Prasad Sharma, Shashi Kant Sinha, Onkar Nath Singh, Ragini, Amita Moitra, Gopal Krishna Kunwar, Prabhat Kumar Roy, Tapan Kumar Ghosh, Sneh Prabha Jha, Mansa Prasad Srivastava, Swapna Chowdhury, Asha Pandey, Anita Pandey, Dhrub Kumar Singh, Syed Shans Tabrez Nasar, Alakhnanda Singh, Manish Chandra Verma, Soma Adhikari, Lalan Kumar Choudhary, Suhasini Besra, Utpala Ghosh, Manoj Kumar Sinha.

I am also grateful to Mrs.Ruma Ghosh Dastidar for her help with the word processor in preparing the final manuscript.

Most of the scientific diagrams illustrated in the book have been drawn by Shri Tribhuwan Poddar.

*Jayashree Datta Munshi*
*Jyotiswarup Datta Munshi*

# Preface

The air-breathing fishes of India have become of great interest in recent years for several reasons. These include the many adaptations of their respiratory and cardiovascular systems for obtaining oxygen from air and/or water, adaptations in their food and feeding habits, reproduction and behaviour. In addition, they provide an important source of protein in regions which are liable to drought and loading.

These fishes are being used for all types of experimental work. Information about their structure and function is scattered in many journals, some of which are limited in availability. Our research work on these fishes has continued for many years. In this book we have endeavored to synthesize for the benefit of students of Zoology, Fisheries and research workers in Biology.

The book on "*Fundamentals of Limnology*" deals with air-breathing fishes which have evolved on several occasions. Evolution of the first bony fishes during the late Silurian and Devonian period produced two new features, lobe fins and the lung fishes.

Animals inhabiting the water/air interface show many morphological and physiological adaptations designed with different types of function in that unique environments. In the present age there are new challenges. These include climatic change and contamination of the environment by anthropogenic activities, Thermal Springs, Lakes and Reservoirs.

The subject matter has been delineated in 11 chapters: Science dealing with biological and other phenomena particularly Inland Waters, Haemopotetic Activity in the Gills of Certain Teleosts, Limnobiotic Studies of the Thermal Springs of Rajgir (Bihar), Natural History of Kosi Basin and The sustainability of hydrological cycle of wetlands of Kosi river basin of north Bihar, India, Reproductive Cycle in *Parreysia Favidens* (Benson): A Freshwater Bivalve of Kosi River, Ponds, Swamps and Marshes,

Diel variations of certain physicc-chemical factors and Plankton population of a *Chaur* (Wetland) of Kusheswarasthan (Bihar, India) and Biological Productivity and Energetics. The book will interest students and research workers in the filed of Limnology.

*Jayashree Datta Munshi*

*Jyotiswarup Datta Munshi*

# Contents

*Acknowledgement*                                                                    *vii*

*Preface*                                                                             *ix*

1. Science Dealing with Biological and other Phenomena
   Particularly Inland Waters                                                          1

2. Haemopotetic Activity in the Gills of Certain Teleosts                              3

3. Limnobiotic Studies of the Thermal Springs of Rajgir (Bihar)                       6

4. Natural History of Kosi River Basin                                               14

5. Reproductive cycle in *Parreysia favidens* (Benson):
   A Freshwater Bivalve of Kosi River                                                33

6. Thermal Springs                                                                   42

7. Lakes and Reservoirs                                                              63

8. Ponds                                                                             84

9. Swamps and Marshes                                                               95

10. Diel Variations of Certain Physico-chemical Factors and Plankton
    Population of a *Chaur* (Wetland) of Kusheswarasthan (Bihar, India)             102

11. Biological Productivity And Energetics                                          108

    Epilogue–Culture of Commercially Important Fishes                               134

    References                                                                      141

    Author Index                                                                    147

    Subject Index                                                                   149

# 1

# Science Dealing with Biological and other Phenomena Particularly Inland Waters

Myself along with my research scholars visited one of the largest wet lands of Darbhanga in North Bihar in the night hours. We found that the air-breathing fishes were very much active in catching insects, which are found below the water hyacinths.

As soon as Sun rises the air-breathing fishes take shelter and sleep under the water hyacinths.

Normally, the amount of oxygen uptake by different biota throughout the twenty four hour cycle is equivalent to the amount of oxygen liberated through the process of photosynthesis during the light regime, *i.e.* day hours. In other words, it may be said that the sum total of the metabolism reflected in the process of oxygen uptake and release of carbon dioxide of different biotic communities is equivalent to the metabolism of the swamp in the utilization of $CO_2$ and liberation of $O_2$ during photosynthesis. It is only when the swamp ecosystem fails to provide the required oxygen, the air-breathing fishes rely more and more on the aerial breathing in order to satisfy their metabolic demand. Thus, it becomes apparent that mainly the dissolved oxygen and the free carbon dioxide govern the rhythm of respiratory metabolism of the fishes living in these swamps. Under normal conditions, in midday hours $DO_2$ is sufficient to sustain basal metabolism of these fishes when they are sleeping under the cover of hydrophytes. When the level of $DO_2$ goes down in the water the fishes become restless and active, and take air-gulps to meet their air-hunger.

Earlier studies on the Indian air-breathing fish species show that most of the species the aerial respiration increases considerably right from the evening hours. In many species, the peaks of the aerial oxygen uptake have been observed during dawn (04-06 h) when the level of dissolved oxygen almost depletes to its minimum. It has been observed that in almost all the Indian air-breathing fish species, the rates of aerial oxygen uptake curiously goes down to its minimum during the mid day hours (12-14 h).

---

Limnology (Gk, Limne, marshy land; Logos – discourse).

# 2
# Haemopotetic Activity in the Gills of Certain Teleosts

The lymphatic and the myeloid organs are the main sources of new blood cells in the vertebrates.

In the brodest sense, the haemopoiesis is in the bone marrow, spleen, thymus, lymph nodes, connective tissue and the vascular and lymphatic systems.

Danchakoff (1916), Doan (1932), Maximoh (1927), Downey (1938) and more recently, Maximoh and Bloom 1960,1964 have come to the conclusion that one primitive multipotential cell, called haemocytoblasts provides the common stem cell for all kinds of blood cells. This has given rise to the Unitarian theory of haemopoiesis against Dualistic or the Trialistic theory of blood formation.

Several haematologists have reported about the haemopoitic activity of the endothelium of the blood vessels. Maximo and Bloom (1960) write that in early embryonic stages, the endothelium of the blood vessels is identical in potentials with the potencies of mesenchymal cells. Thus in the yolk-sack vessels and in the some portion of the aorta of the endothelial cells form clusters of haemocytoblasts.

The endothelial cells of the inter-sinusoidal capillaries in the bone marrow of young animal, marked hypertrophy and hypnoses before the formation of the erythrocytes. Small isolated islands of thromblasts develop from these first stem cells maturing within the capillaries the erythrocytes were said to be carried directly in to circulations.

The present author while studying the micro-circulation of the gills of several freshwater fishes found that the endothelium of afferent blood vessels of certain of these fishes, belonging to the family cyprinidae, show marked haemopiotic activity.

In the first stage of development, the endothelial cells become hyper-trophied and multiphied and multiply to form clusters of cells, hanging freely into the lumen of the blood vessels. Later on, each such clusters becomes detached from the main stream and floats freely in the lumen as a blood island. In *Catla catla* groups of such stream cells – the haemocytoblasts are seen attached to the inner wall of the afferent blood vessel of the gill (Figure 2.1).

Photomicrograph of a section of a gill of *Catla catla* showing groups of haemocytoblasts attached to the inner wall of the afferent blood vessel. X 340. Figure 2.2 Photomicrograph of a section of a gill of *Labeo rohita* showing clusters of haemocytoblasts and blood islands in the lumen of the afferent blood vessel, x 40 Figure 2.3. An enlarges view of Figure 2.2 showing one of the clusters of haemocytoblasts and the blood islets. X 400. Figure 2.4. A highly magnified view of a portion of the above showing the details of the structures of the haemocytoblasts and the blood island. X 100.

In a section of a gill of *Labeo rohita* passing along its afferent vessels, the endothelium of the blood vessel is seen in different stages of haemopoetic activity (Figure 2.2).

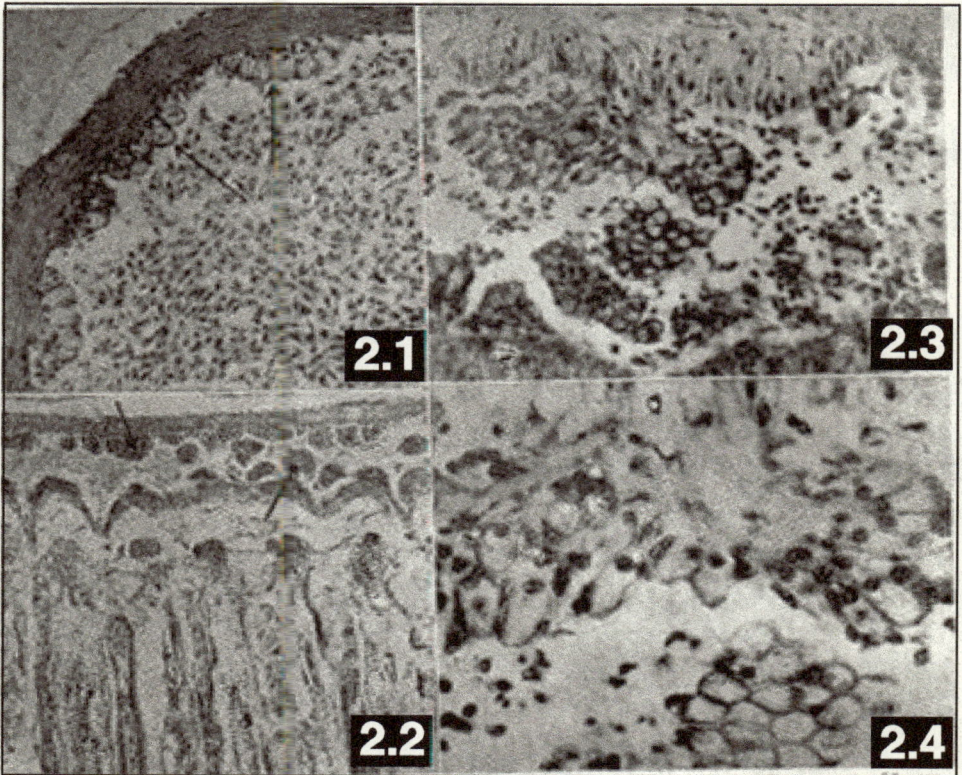

**Figures 2.1–2.4**

The haemocytoblasts developing in clusters from the wall of capillary and project freely into the lumen of the blood vessel and, later on, the bunch of cell, become constricted off from the stream and floats freely as a blood island in the lumen (Figures 2.3 and 2.4) when such an island as cluster of blood cells happens to be cut across it wears a honey-combed appearance (Figure 2.4).

The nuclei take a deep stain with haematoxylin and may lie excentrical. The cytoplasm is slightly eosimophilic and fibrilar in characters.

The haemocytoblasts developing in clusters from the wall project freely into the lumen of the blood vessel and later on the bunch of cells becomes constricted off from the plate as a blood island in the lumen (Figure 2.3).

When such an island or clusters of blood cells happens to cut across it wears a honey-combed appearance. The nuclei take a deep stain with haematoxylin and may lie excentrically. The cytoplasm is slightly osmophilic. These cells will eventually differentiate into future erythrocytes. Here is a case to show that in some freshwater fishes blood cells originate intravascularly from the endothelium of blood vessels belonging to the gills.

# 3
# Limnobiotic Studies of the Thermal Springs of Rajgir (Bihar)

The present paper deals with the limnobiotic studies of the hot springs along thermal gradient located at Rajgir (24° 57′ N Lat: 85°25′ E Long). The temperature remains more or less constant at the source throughout the year. There are variations in the physico-chemical parameters along the thermal gradient, Flora is mainly represented by 6 species of Myxophyceae, 6 species of Chlorophyceae and 10 species of Bacillariophyceae. Fauna is represented by Nematoda, Insects, Annelida, Molluscs and Pisces which are restricted to lower temperature range. The interaction of different physico-chemical parameters undoubtedly controls the distribution of biota in thermal springs.

## Introduction

Existence of living beings in natural hot waters has been attracting the attention of biologists for a period of two centuries. For the first time Sonnereat (1774) reported fish living at a temperature of 82°C in Phillippines. Since then, a large number of workers like Brock (1967), Stockner (1967), Castenholz (1969) and Naiman (1976) have worked on flora and fauna of thermal springs having different temperature gradient. In India, more than 300 hot springs are present. Drouet (1938), Gonzalves (1907), Prasad and Srivastava (1965) and many others studied thermal algae in India. Recently, Saha *et.al* (1978) and Saha and Datta Munshi (1982, 1983) made limnobiotic studies of some of the hot springs of Bihar. However, very little information is available on the thermal springs of Rajgir, Bihar.

With a view to understand the factors influencing the distribution of biota along the thermal gradient of a hot spring, the present work was undertaken. The physico-chemical and biological characteristics along thermal gradient of some hot springs of Rajgir, Bihar (India) form the basis of this paper.

## Study Area

The thermal springs of Rajgir are situated at 24°57'N Lat. And 85°25'E Long. The area is located at 100 km., South East of Patna. There are more than twenty hot springs emerging from the bases of various hills. In Rajgir, there are two types of hot springs:

(a) Cemented enclosed cistern, where thermal water emerges through fissures situated at the bottom of (Brahamkund and Gurunanak Kund) and Projected spouts, through which thermal water comes on and passes either directly (Saptadhara) or passes after storing in cemented cistern (Surya-Kund, Vyas Kund, Makhdum Kund and others). The waters of these hot springs flow and converge to form some main streams. Temperature of the water decreases gradually as it passes away and is used by the farmers.

For the limnobiotic study of thermal springs, five sampling stations (A.B.C.D. and E) were selected in view of variations in their nature and temperature (Figure 3.1). Spots A and B are the two sources of thermal springs known as Saptadhara and

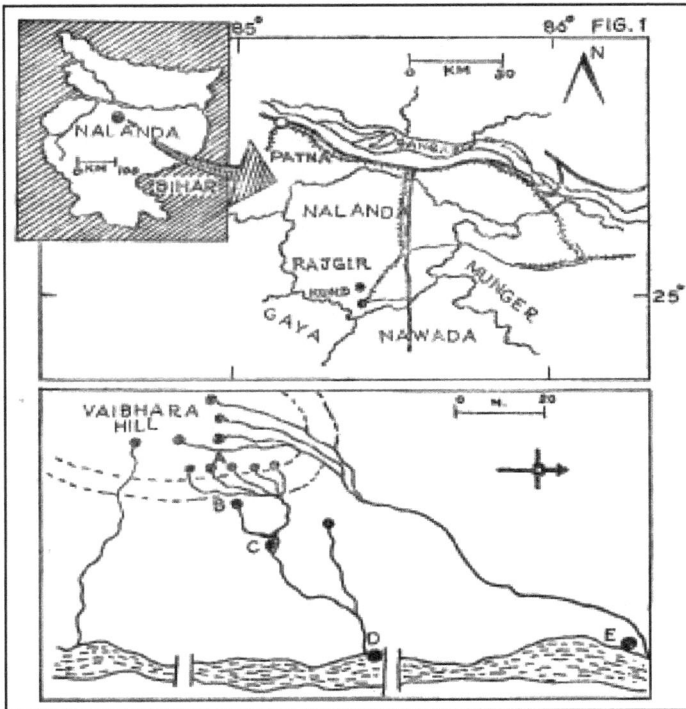

**Figure 3.1: Study Site and Location Map of Hot Springs of Rajgir.**

Brahmkund respectively. Spot C is the joining site of the above two situated at about 15 meters from spot B, where maximum human activities occur. Spot D is the joining site of hot water with a cold stream, situated at about 40 meters from spot C, Spot E is the site of cold water situated at about 70 meters from spot D.

## Materials and Methods

The physico-chemical and biological characteristics of the thermal springs were analysed monthly from March, 1990 to Fabruary 1991. Water temperature, pH dissolved $O_2$, free $CO_2$, Chloride, Carbonate and bicarbonate alkalinity were determined immediately after the collection of the samples. Temperature was recorded with a graduated mercury thermometer. Hydrogen ion-concentration (pH) was determined by a portable Griph pH meter (Systronics type 323, Sl. No.1099). The other chemical variables were determined according to the methods described in the Standard Methods (APHA, 1981). Plankton and other biota collected from different spots were preserved in five percent formalin and their analyses were made in the laboratory later on. Plankton were collected by filtering the water through a plankton net (made up of bolting silk no. 25 with 77 meshes / Sq.Cm.). Samples of periphyton were collected by scraping submerged stones, sticks and other substrata.

## Results

### A) Physico-chemical Parameters

The average values of the physico chemical parameters of various sampling stations for different seasons are presented in (Table 3.1).

The temperature of these thermal springs remains almost constant at source. It is 41-42°C and 41-43°C at spot A and B respectively. As the water flows onward the temperature decreases gradually and at spot E, it varies between 24°C and 30°C in different seasons. The water of the thermal springs is acidic, mainly due to absence of carbonate alkalinity and presence of free $CO_2$ and bicarbonate alkalinity. The water of spot B was found more acidic (pH 5.3-5.8) than that of spot A (pH 6.0-6.2). This is due to more free $CO_2$ at the site B. Free $CO_2$ was also found very high (104 mg/t) at this spot in Summer seasons. The value of free $CO_2$ decreases as the water flows forward and it is very low (18.0-22.0 mg/l) at spot E. The value of bicarbonate alkalinity at spot C and D was observed very high (100-108 mg/l and 174-189 mg/l respectively). It is highest at spot D during Monsoon. The increases in bicarbonate alkalinity and decrease in free $CO_2$ from C to E cause an increase in pH. There is a great difference in the $DO_2$ value of spot A and spot B. The water of spot A contains more dissolved oxygen (6.4-8.5 mg/l) than that of spot B 3.8-6.4 mg/l). It is remarkably higher in Summer at spot A and E. The value of $DO_2$ is quite low at spot C and it increases gradually as the water flows onward and it is maximum (7.0 mg/l) at spot E.

### B) Biota

The biota of the hot springs along the thermal gradient is presented in (Table 3.2). Algae belonging to Myxophyceae, Bacillariophyceae and Chlorophyceae were recorded as in main biota of the hot springs. There are altogether 6 species of Myxophyceae, 6 species of Chlorophyceae and 10 species of Bacillariophyceae.

**Table 3.1: Physico-chemical Factors of Various Sites for different Seasons**

| Physico-chemical Parameters Seasons | Sites | Atm. Temp. (°C) | Water Temp. (°C) | pH | $DO_2$ (mg/l) | Free $CO_2$ (mg/l) | $HCO_2$ (mg/l) | Chloride (mg/l) |
|---|---|---|---|---|---|---|---|---|
| Summer | A | 35 | 42 | 6.1 | 8.5 | 23 | 59 | 7 |
| (From March to June) | B | – | 43 | 5.3 | 6.4 | 104 | 62 | 18 |
| | C | – | 41 | 5.9 | 2.2 | 77 | 100 | 17 |
| | D | – | 35.5 | 6.9 | 3.3 | 38 | 174 | 23 |
| | E | – | 27 | 6.6 | 4.8 | 22 | 120 | 15 |
| Monsoon | A | 30 | 41.5 | 0 | 6.4 | 24 | 56 | 20 |
| (From July to October) | B | – | 42.5 | 5.8 | 3.9 | 97 | 70 | 20 |
| | C | – | 40.2 | 6.1 | 4.1 | 80 | 108 | 25 |
| | D | – | 34.5 | 6.7 | 3.2 | 74 | 189 | 33 |
| | E | – | 28.8 | 6.5 | 7 | 19 | 98 | 18 |
| Winter | A | 23 | 41 | 6.2 | 7.4 | 16 | 66 | 16 |
| (From November to | B | – | 42 | 5.4 | 3.8 | 77 | 76 | 20 |
| February) | C | – | 40 | 5.7 | 2 | 69 | 108 | 26 |
| | D | – | 33 | 6.8 | 3 | 52 | 180 | 30 |
| | E | – | 24 | 6.8 | 6.2 | 18 | 138 | 22 |

## Table 3.2: Biota of Hot Springs of Rajgir at different Spots

| Spots Biota | A Plank. | A Peri | B Plank. | B Peri | C Plank. | C Peri | D Plank. | D Peri | E Plank. | E Peri |
|---|---|---|---|---|---|---|---|---|---|---|
| **ALGAE** | | | | | | | | | | |
| Myxophyceae | | | | | | | | | | |
| Phormidium tenue | + | + | + | + | + | + | + | + | – | + |
| P. fragile | + | + | + | + | + | + | + | – | – | – |
| Oscillatoria supprevis | – | + | – | + | – | – | – | + | – | – |
| O.tenuis | – | + | – | + | + | + | – | + | – | – |
| O.chlorina | – | + | – | + | + | + | – | + | – | – |
| O. raoi | – | + | – | + | – | – | – | + | ▮ | |
| **B. acillariophyceae** | | | | | | | | | | |
| Navicula protracts | – | + | – | – | + | + | + | + | + | + |
| N.cincta | – | + | – | – | + | + | + | + | + | + |
| Pinnularia interrupta | – | + | – | – | + | + | + | + | + | + |
| P.viridis | – | – | – | – | – | – | + | + | – | + |
| Fragilaria intermedia | – | – | – | – | + | + | + | + | – | + |
| Cymbella turgid | – | – | – | – | – | – | + | + | – | + |
| Gomphonema lanceolatum | – | – | – | – | + | + | + | + | + | + |
| G.spheerophorum | – | – | + | – | + | – | + | + | + | + |
| Synedra ulna | – | – | + | – | + | – | + | + | + | – |
| **Chlorophyceae** | | | | | | | | | | |
| Spirogyra sp. | – | – | – | – | + | – | + | – | + | + |
| Oedogonium sp. | – | – | – | – | – | – | – | + | + | + |

Contd...

**Table 3.2—Contd...**

| Spots / Biota | A Plank. | A Peri | B Plank. | B Peri | C Plank. | C Peri | D Plank. | D Peri | E Plank. | E Peri |
|---|---|---|---|---|---|---|---|---|---|---|
| *Ulothrix* sp. | – | – | – | – | – | – | – | + | + | + |
| *Microcystis robusta* | – | – | – | – | + | – | + | – | + | – |
| *Chara* sp. | – | – | – | – | – | – | – | + | – | – |
| *Cladophora* sp. | – | – | – | – | + | – | + | – | – | – |
| **NEMATODA** | | | | | | | | | | |
| *Monunchus macrostoma* | – | – | – | – | – | + | – | – | | |
| **INSECTA** | | | | | | | | | | |
| *Chironomus larva* | – | – | – | – | + | – | + | – | – | – |
| **ANNELIDA** | | | | | | | | | | |
| *Hirudinaria medicinalis* | – | | – | | – | | – | | + | |
| **MOLSUSCA** | | | | | | | | | | |
| *Pila globosa* | – | | – | | – | | – | | + | |
| **PISCES** | | | | | | | | | | |
| *Esomula damicus* | – | | – | | – | | – | | – | |
| *Clarias batrachus* | – | | – | | – | | – | | + | |
| *Channa gachua* | – | | – | | – | | – | | + | |
| *Heteropneustes fossilis* | – | | – | | – | | – | | + | |
| *Mystus tengra* | – | | – | | – | | – | | + | |
| *Puntlus conchonius* | – | | – | | – | | – | | + | |

*Phormidium* sp. and diatoms were found in little quantity at spots A, B and C but, these are abundantly present as periphyton on submerged walls and other substrata.

The waters of spot D and E were rich in various diatoms (*Navicula protracts, Pinnularia Interrupta, Cymbella turgida, Gumphonema lanceolatum, Fragllaria ntermedia* and *Synedra ulna*), blue-green algae (*Phormidium fragile and P. tenue* and *Oscillatoris subbrevis*) and green algae (*Dudogonium* sp., *Spirogyra* sp., *Ulothrix* sp. and *Microcystis robusta*) as plankton as well as periphyton. At spot *E. Cladophora* sp. was observed most frequently as periphyton on submerged rocks and mollusk shels. *Ulothrix* sp. *Oscillatoris* sp. *Osdogonum* sp. and various species of diatoms were found on *cladophora*. Some macrophytes (*Ipomes* sp. and *icchornia crassipes*) form a thick vegetation on the surface of spot D.

Very few zooplankton and animal biota were observed in the water of the thermal springs. However, the periphyton collected from spot C shows the presence of some nematodes (*Monunchus* sp.). In the water of spot D larvae of *Chironomus* and a fish (*Esomus danricus*) were observed. Spot E (cold water) was characterized by 5 species of fishes (*Clarias batrachus, Channa gachua, Heteropneustes fossilis, Mystus tengra and Puntius conchonlus), leech (Hiruduaria medicinalis) and one species of Molluscs (Pila globosa)*.

From spot A to spot B, the Myxophyceae were exclusively present and they show their abundance during Summer and Monsoon. More Bacillariophyceae were recorded at spot D especially during Summer.

## Discussion

The temperature of the water of the hot springs in different seasons remains more or less constant at the source. As the water flows onward, its temperature decreases. The rate of cooling depends on the initial temperature, channel dimensions, volume of water, atmospheric temperature and wind velocity. According to Brues (1928), the water of a thermal spring is characterized by:

a)  Abnormally high and constant temperature;

b)  Presence of certain salts in considerable amount and

c)  Deficiency of dissolved $O_2$ and great excess of $CO_2$ and $H_2S$.

The physico-chemical parameters of the hot springs seem to be interrelated. With the decrease of temperature, there is an increase in dissolved $O_2$ and decrease in free $CO_2$. The decrease in free $CO_2$ and increase in bicarbonate alkalinity results in higher pH. It is very clear from the data that at spot B, free $CO_2$ is very high (77-104 mg/l) and so, pH is comparatively low (5.3-6.8 mh/l). The decrease in free $CO_2$ is due to release of pressure when the spring emerges through the fissures and escape of free $CO_2$ gradually occurs (Saha and Datta Munshi, 1980).

The increase in $DO_2$ along thermal gradient is due to abundance of algae at lower temperature range. The fluctuation of $DO_2$ is well marked at spot E in different seasons. At this spot high $DO_2$ in Monsoon is the result of algal growth and high water current. Low $DO_2$ level at spot C is due to decaying of materials and human activities like bathing and washing. This also results in high bicarbonate alkalinity and chloride level here.

High temperature is characterized by Myxophyceae only while at low temperature ranges, diatoms become the dominant group. So, Myxophyceae is the most important tolerant group of organisms. The temperature range 33–35°C is suitable for the growth of diatoms. As regards the fauna, no species were recorded between the temperature range of 43-40°C. However, Sarkar (1953) recorded a fish at temperature of 40°C. Thus, a number of biotic and abiotic factors interact to control the distribution of biota along the thermal gradient.

# 4
# Natural History of Kosi River Basin

There are more than seventy wetlands (chaurs) spreading all over the Kosi river of North Bihar, India. They are fed by about small channels called Dhars, originating from the Kosi, Burhi Gandak, Baghmati. Balan rivers which are tributaries of the Ganges. Hydrological studies along with the physico-chemical factors of water of Kosi river and Berela Chaur (swamp) of Kusheswarsthan, Darbhanga have been conducted. In summer months, the Kosi river gets water supply from the glaciers, but drastic depletion of water in most of these wetlands occur as they lose their connections with the riverine source. The peripheral regions of the wetlands dry up due to rapid transpiration of water by macrophytes. In the rainy season they get flooded with water of the adjoining rivers. The formation of wetlands and chaurs and their hydrological cycles depend mainly on the annual flush floods of the rivers and partly on the monsoon rain during the period May to September. In this chapter the hydrological features, drainage pattern, water discharge, sediment load, formation of Dhaurs and wetlands and conservation of their biota and their production potentiality, fisheries and the present status of subsidiary industries like, mother of pearl (MOP) button and lime industries have been evaluated. The management strategy for sustainable hydrological cycle of the river basin lies in the construction of a network of canals and reservoirs to control the annual devastating flood and drought. The wetlands and chaurs of North Bihar are some of the most productive aquatic systems of the world, which sustain the life of millions of people.

The financial assistance from the Indian Council of Agricultural Research and the Council of Scientific and Industrial Research, Govt. of India under project No.37/09040/96 are gratefully acknowledged.

*Keywords*: Wetland, Hydrological Cycle, Conservation.

The river Kosi or the 'Kausik of the legends, drains the southern slopes of the Nepal Himalayan terrain (25°20' to 29° 0' N Lat and 85° 20' to 89° 0' E long) sprawled over a catchment area of 70.409 km$^2$ covering the territory of two countries *viz.* Nepal (59.570 km$^2$) and India (10.839 km$^2$, North Bihar). The river originates from the glaciers of the Mount Gasainthan (8.013m) in the form of three main different streams *Viz.*, the Sun Kosi and Arun Kosi and the Tamar Kosi respectively, all joining together at Tribeni – 10 miles upstream off Chatra. In addition to these, the Indravati, Bhote, Likhu and the Dudh Kosi join the Sun Kosi from the North. Thus, this mighty Himalyan river, a combination of seven streams assumes its popular Nepali name, the Sapt Kosi at Tribeni. It has the highest rate of siltation among the river of the world causing land degradation, heavy silt concentration in the run off and formation of wetlands in North Bihar. Due to strong current and heavy sediment load this flushy hill torrent, as it reaches the plains, has oscillated its course over a vast tract of land since 1736. The hydrological features, drainage pattern, water discharge, sediment load, formation of dhars and wetlands and production potentiality of the Kosi river system have also been described.

The river Kosi or the 'Kausiki' of the legends is the wildest and the most notorious and uncertain among the Indian rivers. It drains the southern slopes of the Nepal Himalayas (25° 27' to 29° 0' N lat. and 85° 20' to 89° 0' E long) extending from Gasainthan (8.013m) in the west to Kanchanjunga (8.579m) in the east. With a catchment area of 70.409 km$^2$ covering the territory of two countries *viz.* Nepal (59.570 km$^2$) and India (10,839 km$^2$, North Bihar), the Kosi basin is the third largest (in area) in the whole of India, ranking next to those of the Indus (2.066.655 km$^2$) and the Brahmaputra (2,56,930 km$^2$). Compared to the Ganges length of 2.575 km arising from 3.950m high Gangotri, the river Kosi is only 965 km long, originating from 5.490m high, Tibetan plateau. It also has the highest rate of siltation among the rivers of the world. From the above physical characteristics, it is evident that the Kosi is a violent river in the mountains that has a catchment area too large for its relatively shorter course. The Kosi, due to these topographical and meteorological features, is rated as one of the problematic rivers of the world.

## Derivation of the Name

### 'Kosi': The Kauski or Kosi

The Kosi known for its antiquity has some myths behind it. It is mentioned as Kausiki in epics, such as, the Skandha Puran, Matsyapuran, Balmiki Ramayan, and the Mahabharat, etc. In ancient literature, the river has been described as a large, powerful and sacred stream for holy bath. According to Skandha Puran, Kosi is said to be the daughter of Kausiki Raja, a celebrated Kshatriya King of Gadhi. The legendary semidivine maiden (Nymph) Kausiki had a brother named Vishwamitra, a worshipper of Para Barhma, or the Supreme God. His sister Kausiki, though Kshatriya by birth, was wedded to a Brahamin sage, Richik by name, who had attained perfection by virtue of worshiping gods. The saint later on, got furious with his wife, for his son, unlike his father, rose against worshsiping of gods. Richik cursed his wife and prayed God to convert his wife (Kausik) into a river (vide Buchanan, 1928).

Kausiki, according to another mythicall writing, was a mermaid goddess, a sea woman having the head and body of a lovely lady and her waist ending as the tail of a fish. She was worshipped by the people inhabiting the Matsyadesha, the land of fishes. The Kingdom by that time was delimited between the old beds of Brahmaputra and the Karatoyo to which the Kosi was connected by a water way.

In the epic Balmiki Ramayana, the Kausiki is associated with the Vishwamitra Satyavati episode. Satyavati, the elder sister of the sage, was married to a saint Rehika. She followed her husband to the heaven from where she was transformed into a river, later to be known as the 'Kausiki' and it was on its banks that Vishwamitra built up his hermitage and used to live.

Much like the Ganga, the Yamuna, the Saraswati and the Brahmaputra, the Kosi too has its own religious importance. People of the area take holy bath in this river during festivals. Several temples have been built up on the banks of the river of which the following deserve special mention. – temples of Barrah Mahadeo (8 km downstream of Tribeni), Kala Maik both on the Nepal Hills. Dewanbaba Temple (Nambatta), Karubaba Temple (Mahishi) and Katyansisthan (Salkhua). All these temples are in Sahara District of Bihar.

## Origin of Kosi

The river originates from the glaciers of Mount Gasainthan (8,013m), the Mount Everest (8,848 m) and the Kanchenjunga (8,579m) in the form of three different main streams *viz.*, Sun Kosi, the Arun Kosi and the Tamar Kosi respectively, all confluencing together at Tribeni 10 miles uphill of Chatra town. In the upper catchment basin, the main feeders of Arun are the Phungchu, Bhangchu, Menchu, Shichu, Dzakarchu, Chiblungchu and Yaruchu (chu is a local term for a stream in Tibet.) In addition to these, the Indravati, Bhote, Likhu and the Dudh Kosi join the Sun Kosi from the North. Thus, the Kosi becomes one of the mighty Himalayan rivers comprising of seven streams. It is for this fact that the united stream assumes its popular Nepali name - 'The Sapt Kosi' at Tribeni. As it enters the plains at Chatra (Nepal), it assumes the name of 'Kosi'.

The total catchment of the Kosi, above Chatra in Nepal, is about 59,570 km$^2$ is above the snow line (Table 4.1).

**Table 4.1: Catchments of different Principal Tributaries
Forming the Sapt Kosi at Chatra**

| River | Catchment (km$^2$) | Percentage to total |
|---|---|---|
| Tamar | 5,770.1 | 10 per cent |
| Sun Kosi | 18,985.7 | 32 per cent |
| Arun | 34,524.7 | 58 per cent |
| Sapt. Kosi (Below Tribeni upto Chatra) | 290.0 | Negligible |
| **Total** | **59,570.5** | **100 per cent** |

## Glaciers

A large part of the Kosi catchment (5770 km²) is covered with permanent glaciers of the Himalayas. This accounts for heavy water discharge in the river during the summer. Some of the important glaciers are the Kanchanga (10 miles), Yalung (13 miles). Zemu (16 miles), Renghuck (14 miles) and the Kyebark (11 miles).

## Shifting Courses of the River

The river Kosi is well known for profuse branching into many interlacing channels in its course. It does not remain static in a fixed channel for a long time. Since ancient times, it has been oscillating over a vast tract of North Bihar. This wild movement of the river has neither been steady nor continuous, rather it has been taking sudden jumps from one channel to another. In this process, it deserts the previous ones which still hold on some water (Chibber, 1949), leading to the formation of swamps and wetlands. The only comparable river in this respect is the "Hwang-Ho of China" (Ahmed, 1947).

## Causes of Shifting Tendency

The oscillating character is associated only with the Kosi and no other river of the region. The causes of oscillation have puzzled hydrobiologists, geographers and geologists. Furthermore, the westward shifting tendency of the Kosi as against the natural slope (North-west to South east) and configuration of the region is more amazing. Several explanation given by Das (1968) and Singh (1986) deserves special attention. According to these authors, situation, rapid water discharge and bed slopes of the Kosi flood plain, are responsible for ever-shifting nature of the river.

## History of Shifting Courses

In the early part of the 18th century, the Kosi flowed down the town of Purnea but it has worked west-ward across 120 km upto Saharsa-Darbhanga border. The leftout channel at Purnea is now known as *Kalikosi*. During the period, the Kosi after issuing forth from Chatra turned east and southeast through Purnea and met the Ganga near Manihari and later on its neighbourhood of Karhagola (1856). Since then, the river has been known to have curved out numerous parallel channels during its westward migration from Purnea (1736) to Nirmali (1950). The oscillating shifting nature and its rate of movement is clear.

The Kosi river harbours more than 110 species of teleostean fishes of 66 genera *belonging to 27 families and 10 orders* (Datta Munshi and Srivastava, 1988). Best quality fish spawns of major carps come from the Kosi river during the first flush flood (Jhingran, 1975). The swamps, chaurs and wetlands are the natural abode of about 15 species of air-breathing fishes which are very well adapted to the physico-chemical conditions of these derelict water bodies. Several workshops were held on air-breathing fish culture of which the sixth one was held at CIFRI, Barrackpore, Calcutta in December, 1982 (Datta, Munshi and Choudhary, 1996).

## Production Potentiality of Wetlands/Swamps

### Shell-Fisheries and Mother of Pearl Button Industries

Shell fisheries has been considered as one of the important aspects of the riverine-fisheries of North Bihar, in which such mollusks (as Bivalves and Gastropods) are collected from the Kosi river basin which form raw materials for the mother of pearl (MOP), button industries, and ornaments.

Recently, we have surveyed the different types of aquatic beds of the Kosi river systems to identify various mollusk- collecting centres of North Bihar. Molluscan species so far collected from different aquatic systems of this area have been systematically identified and their commercial importance was evaluated, (Sharma et.al.,1983). Out of 10 species of *Parreysia favidens, P. favidens var. pinax, marcens,* and *Lamellidens corrianus* cater the requirements of MOP button industries. The empty shells after sun dried are processed for the production or lime, poultry feeds, mosaic tiles, fertilizers and chemicals ($CaCO_3$) in these industries.

## Present Status of MOP Button Industries of North Bihar

The shell-fisheries in this belt had a glorious past because of the availability of raw materials in sufficient amount. The number of these industries declined from 3.50 to 1.51 ton by 1940 which declined further to 1.03. Presently only 84 registered units (industries) run in the area which can hardly work for a few months in a year due to less supply of raw materials. These industries now have started manufacturing "Shell-ornaments" of different colours and varieties.

From the bivalve landing data obtained for the period of 1957 to 1971 from different sources such as CSO at Mehsi, we could assess the present status of MOP button industries in this belt. The landing data of bivalves reflect their population structure in terms of biomass in the area.

A histogram of the annual landing data (1957-1971) of bivalves was plotted in which 5 spikes (2+1+2) of bivalves biomass (ton) were discernible at intervals of 2.3 and 4 years in a time span of 1.5 years. The appearance of spikes only at specific intervals of time is difficult to explain. The spikes and the curves (representing the biomass of the shells obtained in clusters of year) obey the general exponential equation $Y=ae^{LR}$. Where a constant (321.23 ton) is a factor indicating together, we had a straight line indicating steep declination of the bivalve population.

## Production of Gastropods and Lime Industries

The gastropod mollusks are eaten by common people, the shell of mollusks are used for manufacturing lime indigenously. Lime so manufactured by this process is consumed with tobacco and betel by local people.

## Integrated Aquaculture

In the country about $0.6 \times 10^6$ ha of water area is supposed to be swampy and unutilisable due to its weed infestation and derelict ecology. The culture of air-breathing fishes like *Channa punctatus* (garai), *Anabas testudineus* (Koi), *Heteropneustes fossilis* (singhi), *Clarias batrachus* (magur), is usually integrated with the raising of aquatic cash crops such as makhana (*Euryale ferox*) or singhara (*Trapa bispinosa*). These operations considerably enhance the income of the farmers. A production of 1200 kg of air-breathing fishes and 3200 kg of Makhana per hectare during a season in Gunsar fish farm Darbhanga (Bihar) has been demonstrated by the local farmers and authorities. Apart from these paddy-cum-fish culture is an age-old practice in these swamps. Vast lowlands are being used for paddy-cum-fish-culture. The paddy which is grown in the marshy lands are hydrophytic in nature Dehadrai, 1982).

The production of makhana (*Euryale ferox*) is produced from one acre of Jalkar (water bodies of different types). As per the recent survey report of the Central Bank of India, about 2,500 such types of Jalkars are located in Darbhanga district alone (Laal, 1981).

This unique freshwater system seems to be highly productive which could be developed further by innovating new eco-developmental strategies with the best utilization of the production potentialities of the area for the upliftment of the socio-economic condition of the inhabitants and for the prosperity of the region.

Studies show that the bio-geochemical cycle operating in the wetland/swamps is more complex owing to its specific pattern of transportation of nutrients. Hydrological pathways include import of nutrients from the surface runoff and discharge from the rivers Kosi, Bagmati and Burhi Gandak. Biological inputs include carbon and nitrogen trapha fixation by benthic autotoplma. Excreta of animals of the offshore region and birds guano add nutrients to the wetlands. This lends to eutrophication of the swamps which are bio-geochemically closed. Due to siltation, eutrophication of wetland/swamps and human habitation surrounding the water systems, there is terrestrialization of the large wetlands and chaurs in the region.

A series of reservoirs and canals are to be constructed in the upper reaches of the rivers for control of flood as well as drought. The dead river channels of the region may be converted into permanent canals to feed the chaurs and wetlands.

During drought season feeding of swamps/chaurs with water is essential. For this a series of reservoirs are to be constructed in the uppear reaches of the Kosi river. The Maradhars (dead Channels) are to be converted into permanent canals for feeding the chaurs.

## The Sustainability of Hydrological Cycle of Wetlands of Kosi River Basin of North Bihar, India

There are more than seventy wetlands (chaurs) spreading all over the Kosi river belt of north Bihar, India. They are fed by about 56 small channels called dhars, originating from the Kosi, Burhi Gandak, Baghmati, Balan rivers which are tributaries of the Ganges. Hydrological studies along with the physico-chemical factors of waters of Kosi river and Berela chaur (swamp) of Kusheswarsthan, Darbhanga have been conducted during the period of June 1986 to May 1988. In summer months, the Kosi river gets water supply from the glaciers, but drastic depletion of water in most of these wetlands occur as they lose there connections with the riverine source. The peripheral regions of the wetlands dry up due to rapid transpirations of water by macrophytes. In the rainy season they get flooded with water of the adjoining rivers.

The formation of wetlands and chaurs and their hydrological cycles depend mainly on the annual flush floods of the rivers and partly on the monsoon rain during the period May to September. In this paper the hydrological features, drainage pattern, water discharge, sediment load, formation of dhars and wetlands and conservation of their biota and their production potentiality, fisheries and the present status of subsidiary industries like, mother of pearl (MOP) button and lime industries have been evaluated. The management strategy for sustainable hydrological cycle of the river basin lies in the construction of a network of canals and reservoirs to control the annual devastating flood and drought. The wetlands and chaurs of north Bihar are some of the most productive aquatic systems of the world, which sustain the life of millions of people.

The river Kosi is one of the wildest and most notorious rivers of the Indian subcontinent. It drains the southern slopes of the Nepal Himalaya (25°20' - 29°0' N; 85°20'-89°20'E) extending from Gasainthan at 8013m altitude in the west to Kanchanjunga 8579m altitude in the east, with a catchment area of 70.409km$^2$ covering the territory of two countries Nepal and India (Figures 4.1-4.3). The Kosi Basin is third largest in area in the whole of India. The river Kosi is only 965km long originating from 5,490 m high Tibetan plateau. It also has the highest rate of siltation among the rivers of the world. From the above physical characteristics, it is evident that the Kosi

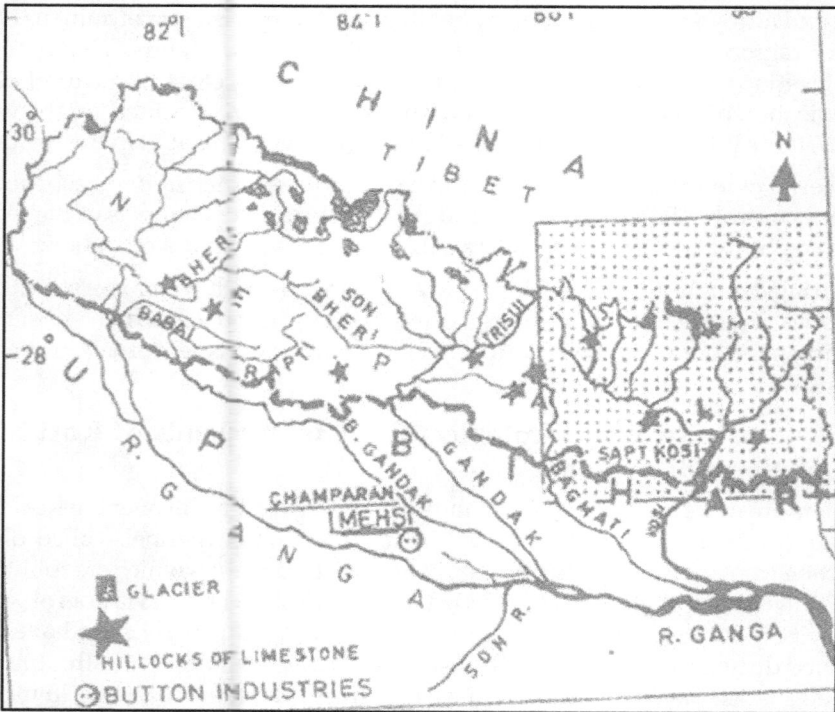

**Figure 4.1: Map Showing Hillocks of Limestone of Nepal and the Origin and Drainage System of River Kosi, Buri Gandak, Gandak, Bagmati and their Confluence with the River Ganga in Indian Subcontinent.**

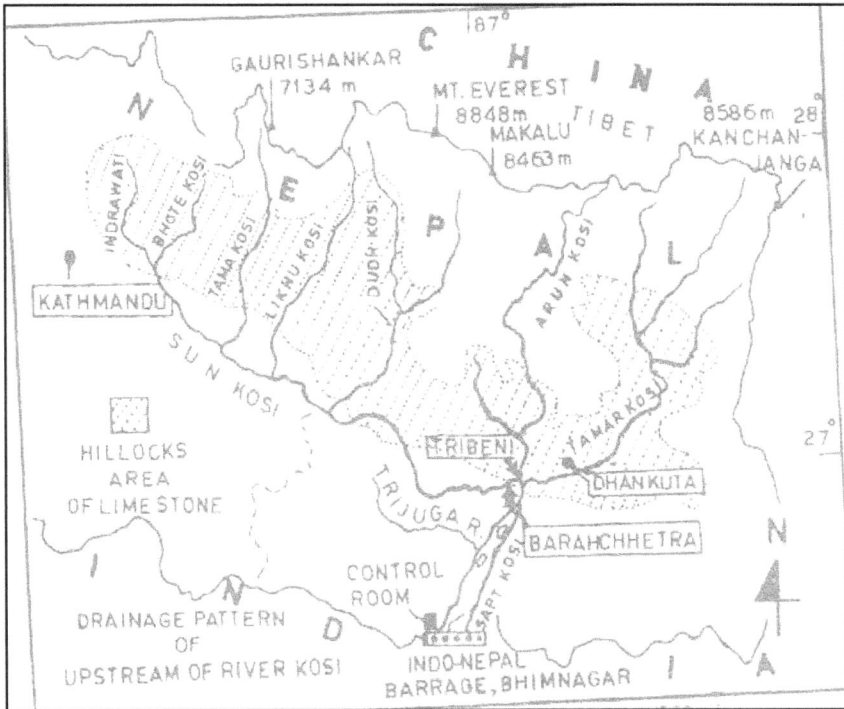

**Figure 4.2: Magnified View of Kosi Basin of Figure 4.1 to Show the Origin of River Kosi from different Altitude and their Drainage System Up to Indo-Nepal Bhimnagar Barrage, Dotted Thatched Line Shows the Hillocks of Limestone.**

is a violent river in the mountains that has a catchment area too large for its relatively shorter course. The Kosi, due to these topographical and meterological features, is rated as one of the problematic rivers of the world. Hydrological studies along with the physio-chemical factors of water of Kosi river and Berela Chaur (swamp) of Kusheswarsthan, Darbhanga have been conducted during the period of June, 1986 to May,1988.

## Materials and Methods

### Hydrological Features

The total water discharge of the Kosi is determined mainly by the glaciers (15 per cent) and rainfall contributing about 85 per cent. Sediment load and siltation has a significant contribution in making of the plains of North Bihar (Chibber, 1949).

### Glaciers

A large part of the Kosi catchment (5770 km$^2$) is covered with permanent glaciers of the Himalayas *e.g.* Kanchanjunga (16.13 km$^2$). Yalung 20.97 km), zemu (25.8 km). Rongbuk (22.58 km) and Kyebark (17.74 km). This accounts for heavy water discharge in the river during the summer.

## Rainfall

The rainfall characteristics of the upper catchment basin of the Kosi deserve special attention. The average annual rainfall for the catchment as a whole, has been calculated to be 1704.3 mm (Jhingran. 1975). In our observations, rainfall data collected from Barrahkshetra (representing the lower catchment area below Tribeni in Nepal) for two years (June, 1986 to May, 1988), indicates that nearly 84 to 96 percent of the total rainfall occurred during the monsoon months (June-October) with an average annual precipitation of 179.3mm and 305.7 mm, respectively. However, maximum precipitation was observed in July 1988 (512.9 mm) and September, 1987 (1090.0 mm) while January was the month of at least rainfall (Nil to 1.9 mm) during 1986-1988.

## Water Discharge

River Kosi is noted for rapidity of its stream and unstable banks. Average discharge in normal years for the Kosi is estimated to be 1,75,000 cusec, although it reached new heights in extremely abnormal conditions during the years 1929 (7,05,000 cusec), 1948 (4,78,442 cusec), 1954 (5,55,000 cusec), 1968 (9,25,000 cusec) and 1984 (4,64,437 cusec). The water discharge data collected from the control room of the Kosi barrage (Bhards, Nepal) for the period, June 1986 to May 1988, revealed that the water discharge of the river increased abruptly with the increase in rainfall during the monsoon season. The average runoff during the period was found to be 83 percent and only 17 percent in rest of the seven months of the year. The peak flow, 1,53,528 cusec and 2,09,082 cusec, was recorded during July 1986 and August 1987, respectively. January, 1987 and February, 1988 were the lean months with 7,005 cusec and 9,247 cusec respectively.

## Physiography and Drainage System of North Bihar

North Bihar plain has an elevation between 150 m in the north west to 25m in the east. It extends from the Himalayan terrain in the north to the Ganga in the south, covering approximately 56980 km². It forms a flat alluvial plain with an average elevation of less than 100 m. The land slopes gently from north east to south west with a gradient of less than 1 m in 5 km. This plain is drained and traversed by numerous river channels and the tributaries of Kosi, Gandak, Burhi Gandak, Bagmati and Mahananda.

The intervening slopes or intercoms have lower gradients and are subject to annual floods. Some of these rivers frequently change their channels.

Due to change of course of the above rivers, especially the Ganga and Kosi many diaras, chaurs and swampy wetlands have come into formation. A large quantity of silt gets deposited in the river beds making them shallow. The entire land of north Bihar plain is dotted with many swampy wetlands (Figure 4.3) (Datta Munshi, *et.al*.1991).

In the region of north Bihar many important wetlands exist of which the Kawar lake (District Begusarai) Gogaheel Chaur (District Katihar) and Kusheswarasthan chaur (District Darbhanga) have been studied.

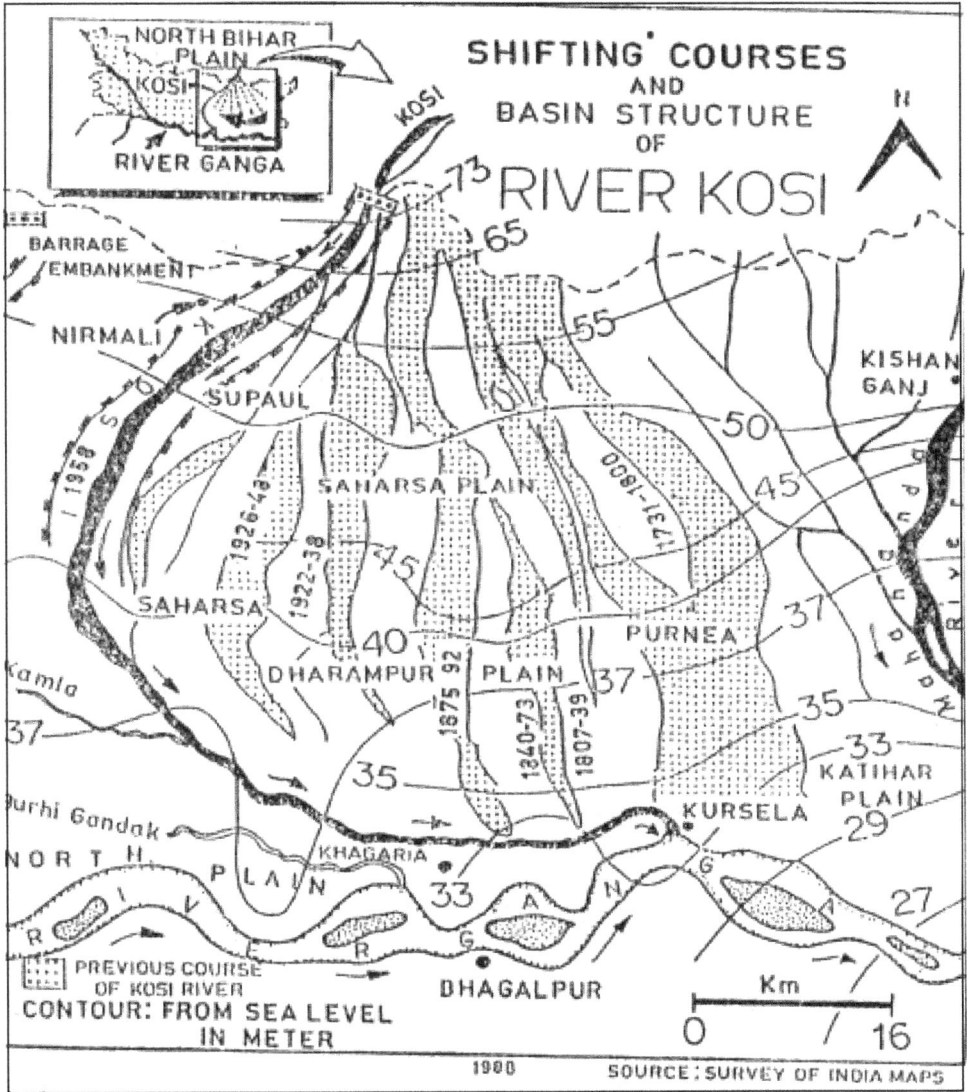

**Figure 4.3: Map Showing Courses of River Kosi and its Basin Structure.**

## Shifting Courses of the River

The river Kosi is well known for profuse branching into many interlacing channels in its course. It does not remain static in a fixed channel for a long time. Since ancient times, it has been oscillating over a vast tract of north Bihar. This wild movement of the river has neither been steady nor continuous, rather it has been taking sudden jumps from one channel to another. In this process, it deserts the previous ones which still hold on some water leading to the formation of wetlands. The causes of oscillation have puzzled hydrobiologists, geographers and geologists.

Furthermore, the westward shifting tendency of the Kosi as against the natural slope (north-west to south-east) and configuration of the region is more amazing. Several explanations have been given by different authorities from time to time. However, the most common and widely accepted explanation is given by Das (1968) and Singh (1986). According to these authors, siltation, rapid water discharge, and bed slopes of the Kosi flood plain, are responsible for ever-shifting nature of the river. In the early part of the eighteenth century, the Kosi flowed down the town of Purnea but it has worked westward across 120 km now known an Kilikosi. During this period, the Kosi after issuing forth from Chatra village turned east and southeast through Purnea and met the Ganga near Manihari and later on in its neighbourhood of Karhagola (1956 A.D.). Since then, the river has been known to have curved out numerous parallel channels during the westward migration from Purnea (1736 AD) to Nirmali (1950 AD). Average percent movement per year shows it to be highest (2.27) during the period 1992-1993. The dynamics of such movement is in all probability dependent besides many other factors, on sediment load of water and sedimentation process in the new channel. However, with the establishment of flood embankments, barrage and irrigation canals, the shifting nature of the river has been greatly contained. It has been bound to flow in between the two Kosi flood embankments traversing partly the districts of Madhubani, Darbhanga, Khagaria and Bhagalpur and Saharas as a whole.

## Physico-chemical Conditions of Kosi River and Swamps

The physico-chemical conditions of the Kosi river have been studied at three sites, one as it enters India at Bhimnagar barrage, second site is at Supaul township and the third one at Kursella, the confluence point of Kosi with the Ganga (Figure 4.3). The data on different parameters of the river water were analysed in three seasons – rainy, winter and summers. The physico-chemical parameters of the water of swamps / chaurs have also been studied.

Physico-chemical conditions of water of Kosi river and water and gas envelop of Kusheshwar Asthan Chaur (Wetland) (The data of physico-chemical parameters of Kosi river are given in Table 4.2.

The physico-chemical conditions of the water of wetland/chaur have been depicted graphically in Figure 4.4. Calcium content of water/soil is considerably high (Table 4.3). This calcium is carried and transported to the Kosi basin by the rivers of Nepal which passes through the calcium hillocks of limestone in Nepal.

## Gas Envelope

A great variety of oxygen conditions exist in Indian swamps. There exists difference in the levels of dissolved respiratory gases in waters infested with free floating, rooted floating or emergent types with combination of submerged vegetation. Generally dissolved oxygen was lower, free carbon dioxide higher, pH lower and temperature fluctuations lower under the water hyacinth areas than in the open water. The extent of difference was found to be dependent upon the time of the day and different seasons (Rai and Datta Munshi, 1979). Due to extreme hypercarbic and hypoxic conditions mostly air-breathers and supplemental air breathers are found living under the coverage of water hyacinth.

**Table 4.2: Average Seasonal Verifications in Physico-chemical Parameters of Kosi at different Sites during July, 1986 to June 1987**

| Parameters | Summer | | | Monsoon | | | Winter | | |
|---|---|---|---|---|---|---|---|---|---|
| | Bhimnagar Barrage (Bipur) Site-I | Supaul (Saharsa) Site-II | Kursela (Katihar) Site-III | Bhimnagar Barrage (Bipur) Site-I | Supaul (Saharsa) Site-II | Kursela (Katihar) Site-III | Bhimnagar Barrage (Bipur) Site-I | Supaul (Saharsa) Site-II | Kursela (Katihar) Site-III |
| Temp. (air) °C | 30.00 | 32.50 | 33.00 | 38.00 | 30.00 | 31.20 | 16.00 | 19.50 | 22.00 |
| Temp.(water) °C | 28.40 | 29.50 | 30.00 | 26.00 | 28.00 | 29.00 | 14.00 | 16.00 | 18.00 |
| Conductivity ($\mu$s-cm$^{-1}$) | 350.00 | 300.00 | 310.00 | 240.00 | 160.00 | 150.00 | 150.00 | 140.00 | 200.00 |
| Turbidity (NTU) | 80.00 | 68.00 | 60.00 | 410.00 | 130.00 | 230.00 | 200.00 | 190.00 | 150.00 |
| T. Solids (mgl$^{-1}$) | 330.00 | 230.00 | 175.00 | 4070.00 | 600.00 | 415.00 | 227.00 | 105.00 | 110.00 |
| pH | 7.00 | 7.30 | 7.50 | 6.90 | 6.80 | 6.70 | 7.50 | 7.20 | 7.40 |
| Free $CO_2$ (mgl$^{-1}$) | 3.00 | 3.20 | 1.70 | 7.00 | 4.60 | 4.00 | 2.80 | 1.60 | 2.00 |
| DO ((mgl$^{-1}$) | 6.20 | 7.00 | 6.60 | 5.36 | 6.52 | 6.00 | 7.20 | 8.00 | 7.90 |
| Total alkalinity (mgl$^{-1}$) | 56.00 | 65.00 | 114.00 | 50.00 | 44.00 | 46.00 | 33.00 | 45.00 | 60.00 |
| Total hardness (mgl$^{-1}$) | 50.00 | 65.00 | 52.00 | 14.00 | 42.00 | 12.00 | 52.00 | 68.00 | 60.00 |
| Calcium hardness (mgl$^{-1}$+) | 46.00 | 52.50 | 44.00 | 10.00 | 28.00 | 4.00 | 50.00 | 60.00 | 50.00 |
| Chloride (mgl$^{-1}$) | 37.00 | 39.70 | 39.70 | 19.50 | 11.11 | 7.94 | 14.20 | 10.54 | 27.80 |
| Silicate (mgl$^{-1}$) | 14.95 | 20.15 | 13.50 | 30.50 | 24.40 | 20.50 | 19.23 | 12.50 | 9.85 |
| Calcium (mgl$^{-1}$) | 18.40 | 31.00 | 17.60 | 4.00 | 11.20 | 1.60 | 20.00 | 24.00 | 20.00 |
| Magnesium (mgl$^{-1}$) | 0.97 | 3.17 | 1.95 | 0.97 | 3.41 | 1.95 | 0.48 | 1.95 | 2.44 |
| Sodium (mgl$^{-1}$) | 3.50 | 5.50 | 6.20 | 3.70 | 6.50 | 3.30 | 1.00 | 1.60 | 2.80 |
| Potassium (mgl$^{-1}$) | 1.80 | 1.00 | 1.40 | 0.80 | 0.50 | 0.60 | 0.00 | 0.30 | 0.00 |
| $PO_4$ –P (mgl$^{-1}$) | 0.02 | 0.10 | 0.00 | 0.00 | 0.33 | 0.00 | 0.02 | 0.05 | 0.04 |
| $NO_3$ –N (mgl$^{-1}$) | 1.70 | 0.23 | 1.00 | 3.06 | 2.33 | 0.80 | 0.00 | 1.80 | 0.80 |

**Table 4.3: Average Seasonal Variation**

| Months | Average Value | |
|---|---|---|
| | Surface | Bottom |
| 1992-93 | | |
| November | 46.86 | 46.99 |
| December | 73.48 | 73.93 |
| January | 76.36 | 76.91 |
| February | 91.68 | 99.47 |
| March | 102.54 | 104.38 |
| April | 105.28 | 104.84 |
| May | 116.25 | 118.55 |
| June | 124.38 | 129.08 |
| July | 84.47 | 87.00 |
| August | 81.86 | 84.90 |
| September | 76.30 | 78.02 |
| October | 71.13 | 71.32 |

In order to clearly illustrate these effects, gas envelopes were constructed for different swamps. Such envelopes are represented graphically by plotting all the monthly measurements of dissolved oxygen and free $CO_2$ within a particular swamp and representing these data with a polygon. These envelopes represented the entire set of concentration of dissolved $O_2$ and free $CO_2$ in the studied swamp and pointed out more clearly that the presence of water hyacinth in polluted swamp created a complete hypoxic and hypercarbic water medium distinctly unsuitable for the development of green algae and most of the water breathing animals. More steep envelop in Makhana (*Euryale ferox*) swamps was evident but it was flatter than the gas envelope of *Chaur* swamps which was infested with emergent *Cyperus* reed plants with a combination of sufficient submerged macrophytes. There was comparatively more dissolved oxygen and less free $CO_2$ concentration in *Chaur* swamps than Makhana and DMCH (Darbhanga Medical College Hospital) swamps (Rai, 1980) (Figure 4.4 a-b).

## Biotic Component

The biotic component of swampy environments is diverse and varied. These mainly include macrophytes, phytoplankton, zooplankton, periphyton and macroinvertebrates and air-breathing fishes.

### Macrophytes

Four characteristic vegetarians communities were recognized in swamps. These are (i) Free floating type *e.g., Eichhornia crassipes, Trapa bispinosa*, (ii) Bottom rooted floating type *e.g., Euryale ferox*, (iii) Submerged type *e.g., Hydrilla verticillata, Potamogeton crispus, Najas graminea, Ceratophyllum demersum* etc. and (iv) Emergent type *e.g., Cyperus* spp.

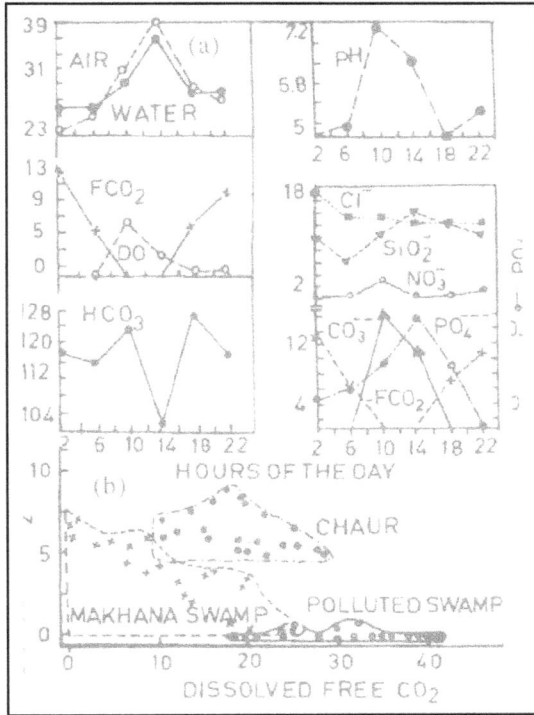

**Figure 4.4: (a) Diurnal Variation of Physico-chemical Factors of Swamps and Wetland/Chaurs, (b) Graph Showing the Nature of Gas Envelops of Three different Types of Swamps.**

There is general agreement that the freshwater swamps have higher macrophytic productivities than any other natural community. Productivity of submerged macrophytes, though considerably high, are always at stressed condition in such habitat, where large floating and emergent aquatic weeds grow. The submerged vegetation was most sensitive to physico-chemical conditions. Mandal (1988) has shown in Laboratory experiments how much Oxygen (mg/l) is consumed in decomposition of aquatic plants (Table 4.2 and 4.3). The problems due to infestation of aquatic weeds are great as they affect the general physico-chemical and biological properties of swamps. However, recently these aquatic weeds are becoming important in the treatment of polluted waters. They absorb potential harmful heavy metals like calcium, nickel, and mercury. The scientists are recommending the use of water hyacinth to remove nutrients from wastewater prior to its discharge into the river.

**Phytoplankton**

The phytoplankton communities of swamps are mainly represented by three groups of algae; Cyanophyceae (Blue green algae), Chlorophyceae (Green algae) and Bacillariophyceae (diatoms) (Datta-Munshi, 1993).Usually the phytoplankton were highest during winter and early summer, while lowest during monsoon seasons.

Billariphyceae dominated among the phytoplankton communities throughout the year. In North Bihar swamps usually two seasonal peaks of diatoms were found. The primary peak prevailed from winter to early summer while secondary peak started developing in monsoon reaching its maximum in winter.

### Zooplankton

Generally zooplankton communities of swamps belonged to Rotifera, Cladocera and Copepoda. The rotiferan population always dominate over other zooplankton communities in swamps. North Bihar swamps are characterized by rotifer genera like *Brachionus, Filinia, Keratella* and *Polyarthra.*

### Periphyton

The periphyton communities of the swamps mainly include algae and testaceous rhizopods (*e.g., Euglypha, Difflugia, Centropyxis, Arcella* etc.). Usually the periphytic species prefer rich $O_2$, high pH and more alkaline medium. Hard testaceous shells of rhizopods are mainly composed of fine sand particles of silica and decomposed diatom cells. The abundance of Bacillariophyceae and silicate contents of water thus control their numbers in a testaceus rhizopod population.

### Macroinvertebrates

Generally the macroinvertebrates of swamps belonged to Annelida, Gastropoda, Odonata, Ephemeroptera, Diptera, Hemiptera and Coleoptera. The aquatic insects and Gastropods dominate the weed-infested swamps. The shallow standing water with macrophyte infact provide a variety of niches for several insect larvae adapted for boring into the stems and leaves of plants and substrate for benthos. Floating and submerged macrophytes seemed to support larger invertebrate fauna than emergent vegetation.

The fluctuation in the density of macro-invertebrate population depend upon various factors:

    i) Stages of life cycle of particular group;

    i) Water depths and vegetation conditions that control directly the physico-chemical conditions of water;

    ii) Different relationship existing between biotic components in the form of interspecific and intraspecific competition and prey-predator interaction and animal-plant relationships.

Depending upon the degree of associations of macroinvertebrate with aquatic macrophytes, they can easily be differentiated into two major groups; (a) The fauna closely associated with submerged macrophytic parts of vegetation (annelids, chironomids, odonata and ephemeroptera) and (b) others comparatively less associated or generally free moving types (Gastropoda, Hemitera and Coleoptera). These associations depend upon different relationships existing between biology of both plants and animals.

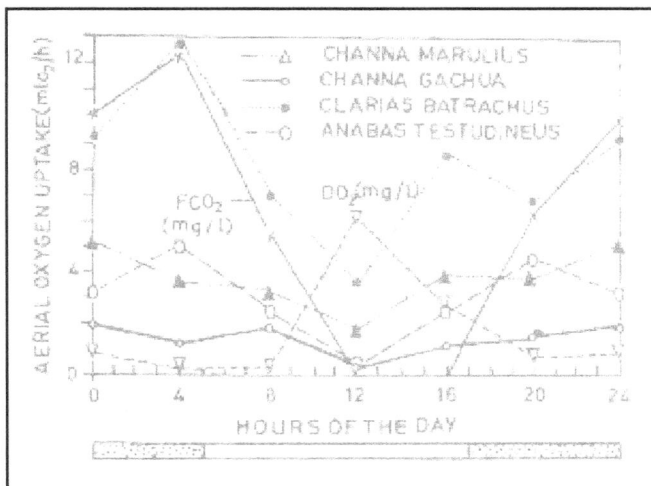

**Figure 4.5: Graph Showing the Relationship between Aerial Oxygen Uptake and Hours of the Day in some Air-breathing Fishes.**

## Fishes and Fisheries

The Kosi River harbours more than 110 species of teleostean fishes of 66 genera belonging to 27 families and 10 orders (Datta Munshi and Srivastava, 1988). Best quality fish spawns of major carps come from the Kosi river during the first flush flood (Jhingran, 1975). The swamps, chaurs and wetlands are the natural abode of about 15 species of air-breathig fishes which are very well adapted to the physico-chemical conditions (Figure 4.5) of these derelict water bodies. Several workshops were held on air-breathing fish culture of which the sixth one was held at CIFRI, Barrackpore, Calcutta in December,1982 (Datta Munshi and Choudhary, 1996).

## Production Potentiality of Wetlands/Swamps

### Shell-Fisheries and Mother of Pearl Button Industries

Shell-fisheries has been considered as one of the important aspects of the riverine-fisheries of north Bihar, in which such mollusks (as Bivalves and Gastropods) are collected from the Kosi river basin which form raw materials for the mother of pearl (MOP), button industries, and ornaments.

Recently, we have surveyed the different types of aquatic beds of the Kosi river system to identify various mollusk collecting centres of north Bihar. Molluscan species so far collected from different aquatic systems of this area have been systematically identified and their commercial importance was evaluated, (Sharma *et.al.* 1983). Out of 19 species of *Parreysia favidens*, *P. favidens* var pinax, *marcens,* and *Lamellidens corrianus* cater the requirements of MOP button industries. The empty shells after sun dried are processed for the production or Lime, poultry feeds, mosaic tiles, fertilizers and chemicals ($CaCO_3$) in these industries.

## Present Status of MOP Button Industries of North Bihar

The shell-fisheries in this belt had a glorious past because of the availability of raw materials in sufficient amount. The number of these industries declined from 350 to 151 by 1940 which declined further to 103. Presently only 84 registered units (industries) run in the area which can hardly work for a few months in a year due to less supply of raw materials. These industries now have started manufacturing "Shell-ornaments" of different colours and varieties.

From the bivalve landing data obtained for the period of 1957 to 1971 from different source such as CSO at Mehsi, we could assess the present status of MOP button industries in this belt. The landing data of bivalves reflect their population structure in terms of biomass in the area.

A histogram of the annual landing data (1957-1971) of bivalves was plotted in which 5 spikes (2+1+2) of bivalve biomass (ton) were discernible at intervals of 2,3 and 4 years in a time span of 15 years (Figure 4.6). The appearance of spikes only as specific intervals of time is difficult to explain. The spikes and the curves (representing the biomass of the shells obtained in clusters of year) obey the general exponential equation $Y=ae^{L.K}$. Where is a constant (321.23 ton) L is a factor indicating depletion rate of bivalve population (L=2.73) and k is the intervals of time in years. Furthermore, when the peak values of annual biomass production of the bivalves in different periods were considered together, we had a straight line indicating steep declination of the bivalve population.

## Production of Gastropods and Lime Industries

The gastropod mollusks are eaten by common people, the shell of mollusks are used for manufacturing lime indigenously. Lime so manufactured by this process is consumed with tobacco and betel by local people.

## Integrated Aquaculture

In the country about $0.6 \times 10^6$ ha of water area is supposed to be swamy and unutilisable due to its weed infestation and derelict ecology. The culture of air-

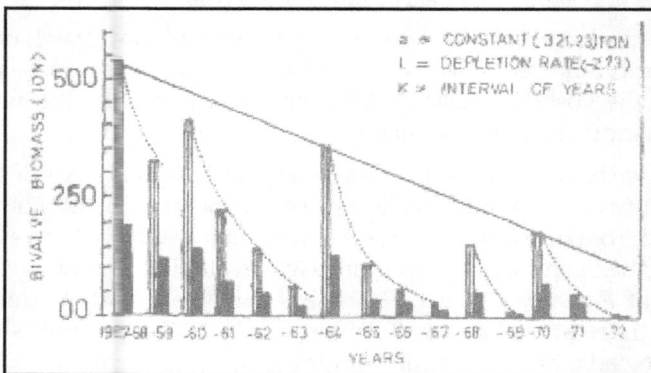

**Figure 4.6: Graph Depicting the Exponential Equation $Y=ae\,L^k$ of Bivalve Shell and MOP Button Production during the Period 1957-1972.**

breathing fishes like *Channa punctatus* (garai), *Anabas testudineus* (Koi), *Heteropneustes fossilis* (singhi), *Clarias batrachus* (magur), is usually integrated with the raising of aquatic cash crops much as makhana (*Euryale ferox*) or singhara (*Trapa bispinosa*). These operations considerably enhance the income of the farmers. A production of 1200 kg of air-breathing fishes and 3200 kg of Makhana per hectare during a season in Gunsar fish farm, Darbhanga (Bihar) has been demonstrated by the local farmers and authorities. Apart from these, paddy-cum-fish culture is an age-old practice in these swamps. Vast lowlands are being used for paddy-cum-fish culture. The paddy which is grown in the marshy lands are hydrophytic in nature (Table 4.3) (Dehadrai, 1982).

**Table 4.4: Showing Comparative Yield of some of the Most Important Commodities of Swamps**

| Sl.No. | Agricultural Commodities | Local Name | Yield Quantities $(ha^{-1}/yr^{-1})$ |
|--------|--------------------------|------------|-----------------|
| 1. | Paddy variety BR. 144 | Charha | 20-25 |
| 2. | Paddy variety BR 46 | Dhan | 20-25 |
| 3. | Paddy variety BR 46-117 | Jasria | 25-30 |
| 4. | Panidhan or Pankaj | Panidhan | 45-50 |
| 5. | *Euryale ferox* | Makhana | 10-12 |
| 6. | *Trapa bispinosa* | Singhara | 04-06 |

The production of makhana (*Euryale ferox*) is produced from one acre of Jjalkar (water bodies of different types). As per the recent survey report of the Central Bank of India, about 2500 such types of Jalkars are located in Darbhanga district alone (Laal, 1981).

## Conclusion

This unique freshwater system seems to be highly productive which could be developed futher by innovating new ecodevelopmental strategies with the best utilization of the production potentialities of the area for the upliftment of the socioeconomic condition of the inhabitants and for the prosperity of the region.

Studies show that the biogeochemical cycle operating in the wetland/swamps is more complex owing to its specific pattern of transportation of nutrients. Hydrologic pathways include import of nutrients from the surface runoff and discharge from the river Kosi, Bagmati and Burhi Gandak in monsoon. Biologic inputs include carbon and nitrogen fixation by benthic autotrophs. Excreta of animals of the offshore region and of birds *guano* add nutrients to the wetlands. These lands too eutrophication of the swamps which are biogeohemically closed. Due to siltation, eutrophication of wetland/swamps and human habitation surrounding the water systems, there is terrestrialization of the large wetlands and chaurs in the region.

A series of reservoirs and canals are to be constructed in the upper reaches of the rivers for control of flood as well as drought. The dead river channels of the region may be converted into permanent canals to feed the chaurs and wetlands.

During drought season feeding of swamps/chaurs with water is essential. For this a series of reservoirs are to be constructed in the upper reaches of the Kosi reiver. The 'Maradhars' (Dieying Channels) are to converted into permanent canals for feeding the chaurs.

# 5

# Reproductive cycle in *Parreysia favidens* (Benson): A Freshwater Bivalve of Kosi River

## Introduction

Thefreshwater bivalves *Parreysia favidens* forms as important shell fisheries of India. The freshwater mussel *Parreysia favidens* (Benson) belong to class Bivalvis, family Unionidae. It is widely distributed in Freshwater. They are found dhars and channels of river Kosi in North Bihar. They live in river bottom on sand, mud, with their ventral part of the shell valves remaining exposed. In the present case a detailed study was done on reproductive cycle, fecundity and development of gonads, particularly ovaries of *Parreysia favidens* collected from Burhi Gandak and dhars of Kosi river of North Bihar.

Sex in bivalves is a subject of great interest. They are either dissolves or monoceious. In some of them change of sex reversal also takes place. A perusal of literature reveals that the work on sex and seasonal gonadal changes has been done mainly with regard to oyster and a few other polycypods. There is little information on *Parreysia favidens*. Patil and Pal (1967) have worked on seasonal gonadal changes in adult freshwater mussel, *Parreysia* var. *marcens* collected from the Mule river, Kirkee (Poona). Patil and Bal (1976) also studied the seasonal changes in chemical composition of the freshwater mussel *Parreysia favidens* var. *marcens* (Benson). Most of the studies on sex and seasonal gonadal changes have been made in relation to oyster *Venus marcensrie*, Bloomer (1930,1931,1934,1935 and 1939 made observation on *Longillidens margindie* and *Anadomers eygenea* in respect of sex and gonadal changes. Agrawal (1980) made some collections from ovarighat located near Jabalpur and studied the seasonal variation in the gonads of *Indonaia caerdea* (Lea).

Bloomer (1930) mace note on the sex of *Anadonia cygnea* collected from Bracabridge pool, satton Park. Italyand Lester (1991) worked on sperm ultrastructure in the Australian oyster *Saccoeires commerclattis* (Iredale and Ronglley) collected from rocks and jelly pylons at Donnybrook (Pulmicestone Passage) southern Queensland Baver (1987) worked on reproductive strategy of the freshwater pearl mussel *Mungerlitfera margarlitfera*. According to the available literature (Hendejberg 1961, Smith, 1979) the perl mussel was reported to be dioecious. Morton (1982, 1986) worked on some aspects of the populations structure and secual strategy of *Corbicula fluminallis* (Bivalves; Corbicullacea) from pearl river of Peoples Republic of China.

In the present study, the reproductive capacity, fecundity and development of gonads, particularly ovaries of *Parreysia fevidens*, collected from Burhi Gandak of Siuri Ghat of Kosi river of North Bihar have been worked out.

## Materials and Methods

The live specimens of *Parreysia favidens* were collected by Ekama dredges from the river Burhi Gandak at Siuri Ghat from 3° to 5° depth in water and kept in aquaria containing sand and tap water in the laboratory for a fortnight. They were fed with live plankton twice in a week. For experimental purpose, the live animals were exposed in normal saline and the ovarian tissue was dissected out and fixed at aceto-alcohol overnight and then transferred to 70 per cent alcohol. After fixation a small portion of the gonadal tissue was taken on a slide. If the tissue became hard and stiff the tissue was dipped into N/10 HCl for 20 to 30 seconds. After this the tissue was teased and isolated from the connective tissue and other interstial substances that might have been present. The tissue was taken and put into the acetocarmine solution.

The macerated tissue was placed on a slide and covered with a thin glass cover-slip and was gently tapped with finger tips. The germ cells of the ovary got segregated.The margins of the cover slip was then sealed by nail polish or sealing wax.

The squash preparation of ovaries was the studied under light microscope, the diameter of ova was measured with the help of ocular and stage micrometers (in μm) under high magnification. There were many ova of different sizes.

Fecundity: The number of ova or eggs present in the ovary of one species or individual is called fecundity. In other words, fecundity (i) the power of individual to multiply, (ii) capacity to form reproductive element; (iii) the number of eggs produced by an individual or a species.

A total of the live animals were used. The ovaries with the mesentries of each animal were dissected out. The weight of the ovaries mass was taken and discharged into 3ml of distilled water. The suspension was made homogeneous and sieved through a metallic net or mesh (size 160 μm) to exclude the unwanted tissues. A pinch of eosin powder was added to stain the ova present in the suspension. The stained suspension was taken in RBC counting chamber of slide with the help of a pipette and the counting was done at 400 fold magnification. For each animal the counting ova was made in 5 small squares on a diagonal line for 5-times (RBC counting chamber has 25 small squares). The ova gave an average count of ova in 0.1 mm$^2$ volume (0.1 µl) of media. Mean count was obtained from average counts from 6 animals.

The data generated from these counts were arranged in three groups and then averaged (group average). The average value being the mean count of ova present in 1 ml of preparation whereas the sum of three groups was the total number of ova present in 3ml of suspension of the whole ovary. From the mean value of ova of all animals the average value of mean count of ova/animal was determined.

## The Sexes are Separate in *P. favidens*

The smallest sized ova were found in the month of August to December, 0.0-75 µm (about 94 per cent). During the months of April to July, the largest size of ova were encountered (Figure 5.1). The results were compilled and the percentage frequencies of different size categories were tabulated. The values vary in different months of the year (Table 5.1).

With gradual increase in temperature of the atmosphere and water during February and March, the ova gained in size and were in the size range of 0.75 and 1.50 to 2.25 µm diameter. During the month of April the smallest size of ova found were; 0-075 µm 33 per cent, 1.50 µm 43 per cent, 2.25 µm 16 per cent and the 3.00 µm 8 per cent. With gradual increase in temperature of the atmosphere as well as water in the month of May, the ova of different sizes were observed *viz.* 0-0.75 µm (1.0 per cent), 1.50 µm (10 per cent), 2.25 µm (20 per cent), 3.00 µm (30 per cent) and the 3.75 µm (19 per cent), 4.50 µm (11 per cent) and 5.25 µm (9 per cent) the largest size of ova.

During the months of May and July the ova start increasing in their volume. Therefore, the average frequency of enlarged size ova was evident in May and July. This increase in volume was quite naturally due to the accumulation of yolk, fat, etc. in the cytoplasm of matured ova. The sizes observed in June were 0.075 µm (23 per cent), 1.50 µm (19 per cent), 2.25 µm (17 per cent), 3.00 µm (15 per cent), 3.75 µm (10 per cent), 4.50 µm (11 per cent) and the 5.25 µm (5 per cent). In the month of July the ova were observed to be 0.075 µm (38 per cent), 1.50 µm (11 per cent), 2.25 µm (12 per cent), 3.00 µm (13 per cent), 3.75 µm (10 per cent), 4.50 µm (10 per cent) and the 5.25 µm (6 per cent). After the release of the matured ova by the end of May and July, the ovary was left out with only small sizes ova (stem cells) (Table 5.1).

Like all germ tissues in the present investigation, there as stem cell population which proliferates to produce a new crop of stem cells plus the maturing ova. As such, the smallest sized ova were present during the whole year which represent the stem-cell population. The maturing intermediate cells (Oogonia, primary oocyte, secondary oocyte etc.) were variable during April to July. By the end of July the mature cells were ready to be fertilized (Figure 5.1).

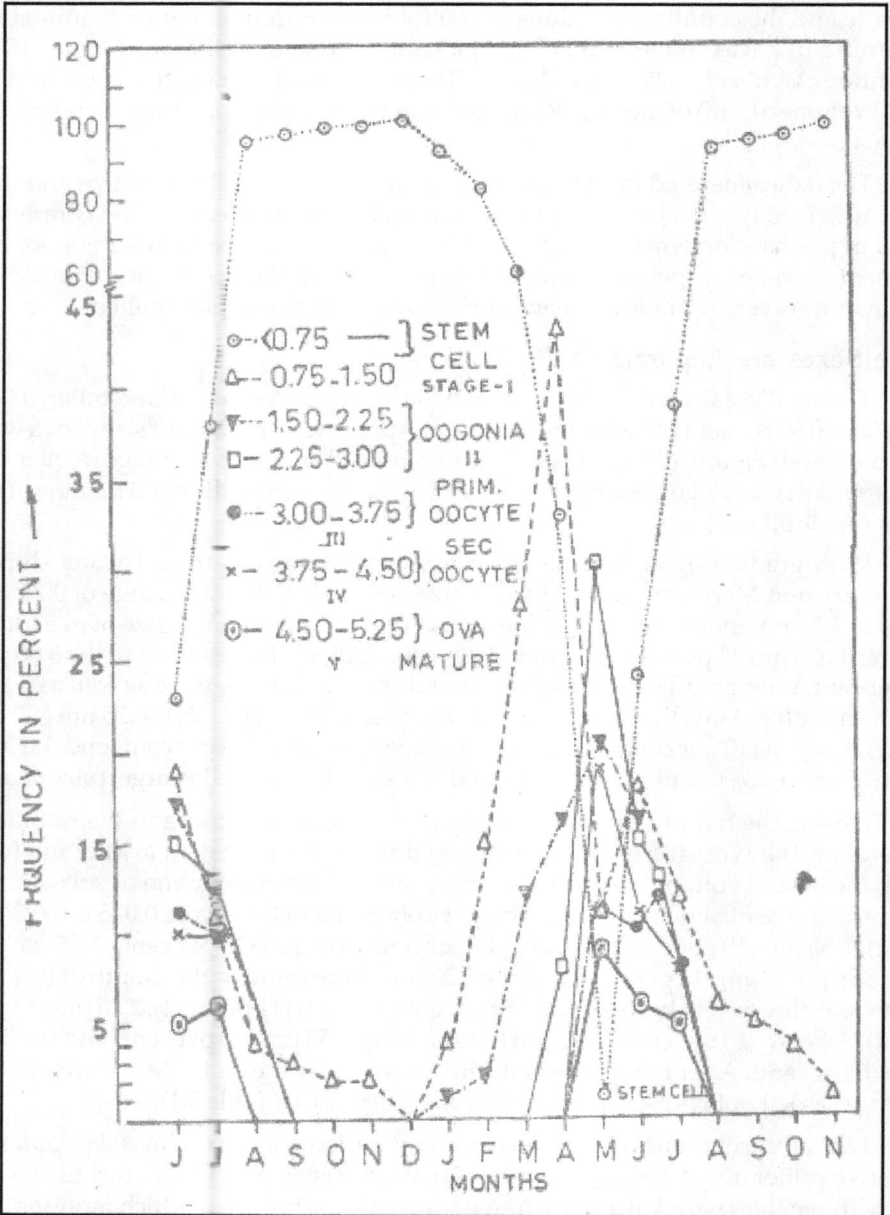

**Figure 5.1: Graph Showing Monthly Variation is the Frequency of different Size Groups of *Parreysia favidens* during June'93–November, 1994.**

**Table 5.1: Ova Frequency of different Size as Present in the Ovaries of *Parreysia favidens* in different Months during One and Half Year (June, 1993–Nov. 1994)**

| Stage | Diameter of Ovum | Jun | July | Aug | Sep | Oct | Nov | Dec | Jan | Feb | Mar | Apr | May | Jun | July | Aug | Sep | Oct | Nov |
|---|---|---|---|---|---|---|---|---|---|---|---|---|---|---|---|---|---|---|---|
| Stem cell I | 0-0.75 | 23 | 38 | 96 | 97 | 98 | 98 | 100 | 94 | 83 | 60 | 33 | 1 | 24 | 39 | 94 | 95 | 96 | 99 |
| | 0.75-1.50 | 19 | 11 | 4 | 3 | 2 | 2 | – | 5 | 15 | 28 | 43 | 10 | 18 | 12 | 6 | 5 | 4 | 1 |
| Oogonia II | 1.50-2.25 | 17 | 12 | – | – | – | – | – | 1 | 2 | 12 | 16 | 20 | 16 | 9 | – | – | – | – |
| | 2.25-3.00 | 15 | 13 | – | – | – | – | – | – | – | – | 8 | 30 | 15 | 13 | 43 | – | – | – |
| Pri, Oocyte III | 3.00-3.75 | 10 | 10 | – | – | – | – | – | – | – | – | – | 19 | 11 | 10 | – | – | – | – |
| Sec.Oocyte IV | 3.75-4.50 | 11 | 10 | – | – | – | – | | | | | | 11 | 10 | 12 | | | | |
| Ova Mature V | 4.50-5.25 | 9 | 6 | – | – | – | – | | | | | | 9 | 6 | 5 | | | | |
| | | 100 | 100 | 100 | 100 | 100 | 100 | 100 | 100 | 100 | 100 | 100 | 100 | 100 | 100 | 100 | 100 | 100 | 100 |

## Description of the Morphology of Ova at Different Stages of maturity

Five stages of development of cva were observed (Plate 5.1, Figure 5.1-5.6)

### Stem Cells

The stem cell population is present throughout the year. Upon division it produces two daughter cells, one of which remains as stem cell. The other daughter cell undergoes differentiation to produce mature ova after passing through intermediate stages of oogenesis.

The stem cells were observed as very-minute bodies poorly distinct nucleus and scanty cytoplasm. Usually circular in outline, but sometimes ova or ellipodal stem cells were also observed.

### Intermediate I Stage

The ovum of this stage was very small. It is surrounded by vitelline membrane. The cytoplasm is almost clear with small yolk granules and a nucleus.

### Intermediate II Stage

The ova of this stage appeared to have an external semi-transparent vitelline membrane of slightly wavy mature. This coat is separated from the egg by perivitelline space. Cytoplasm of the ovum is full of yolk particles. The anterior region of the ovum had a poorly distinct nucleus with chromatin network. Dispersed in the internal "medullary" portion of the ovum were seen a few yolk/fat granules and droplets.

The "cortical region" of the stage II was not traceable. It was quite likely that the "cortical region" might be temporary stage of the raw materials which was consumed during vitellogenesis.

### Intermediate III Stage

It represents a dividing stage of ovum with perivitelline membrane. A nucleus was often present in the centre of the cell which was circular in outline. Very small yolk particles in granular form were fairly present. Bifurcation of the shell takes place at this stage.

### Intermediate IV Stage

In this stage the mature ovum is seen with rod-shaped particles and yolk granules. The egg is surrounded by a semi-transparent gelatinous sheath of slightly fibrous mature.

### Stage V

The large sized end cells were encountered in cell population.

A full mature ovum was fairly of large size and was developing into a larva with shell and adductor muscle. A well developed glochidium larva is surrounded by shell valves with adductor muscle and byssus thread.

April to July in the breeding season of *P. favidens*. In the last week of July one to two size groups of the matured ova are discharged. During this period the gills

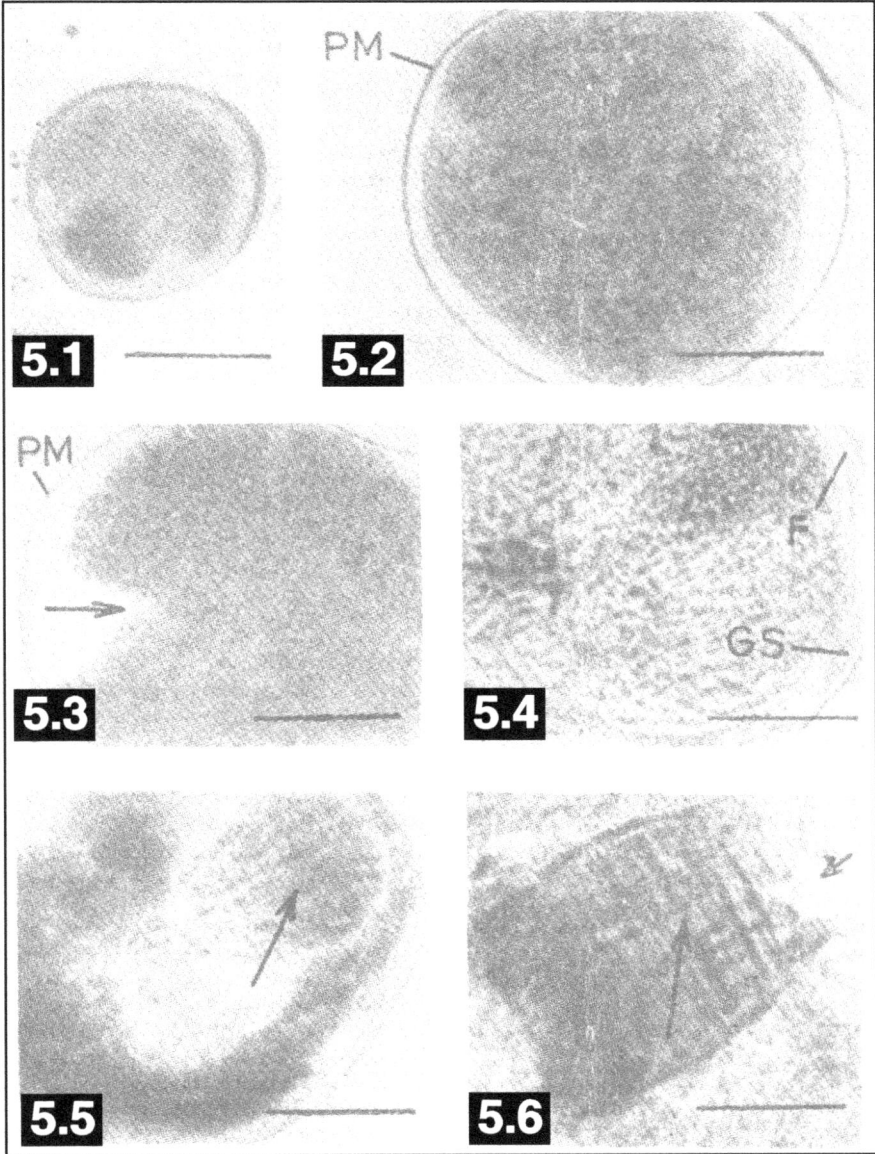

**Plate 5.1**

**Figure 5.1: Fertilised ovum with nucleus, viteline membrane and very small yolk granules. 2: Larger stage of ovum with cytoplasm full of yolk particles enclosed within periviteline membrane (P.M), 40x4; 3:Dividing stage (arrow) of ovum with periviteline membrane, 40x4. 4: Showing mature ovum with condensed fibrilleae and yolk granules. The egg is surrounded by a semi-transparent gelatinous sheath (GS) of fibrous (F) nature. 40 x 4. 5: Small developing larva with arrow shell and adductor muscle, 40 x 4, 6: Developing glochidium larva with byssus thread (BT) (arrow) and shell with adductor muscles (arrow). 40x4. Bar = 500µm.**

become thickened, hundreds of eggs are discharged in the mantle cavity of the body and are carried through the ostia into water-tubes of the outer and Inner gill lamina.

They are held by the mucous secretion of gills and are then fertilized by sperms. The fertilized eggs are retained in the gills which develop into the glochidium larva. The outer and inner gill lamina get greatly enlarged due to the accumulation of a large number of embtyos. The gills act as breeding pouches "marsupial". For housing the fertilized eggs remarkable changes take place in the gills structure (Munshi and Pandey, 1991). Females are easily recognized at this stage by the swollen appearance of their outer gill plates.

Glochidia remain entagled by their "byssus thread and are nourished with the mucous secreted by the gills which act as brood pouches. When sufficiently mature, they come out through exhalent siphon into the water where they slowly sink to the bottom or get scattered by water current.

The data generated from counts were arranged in three groups and then averaged (group averaged); the average value is the mean count of ova present in 1 ml of preparation, whereas the sum of three groups is the total number of ova present in 3 ml of suspension of the whole ovary. Then from the results of combined mean values of all animals the average of ova produced by one individual was computed.

The total count of ova produced per animal was 3568333 S.E. ± 1.31694 (Table 5.2).

**Table 5.2: Distribution of Ova Count in the Ovary Suspension (3ml Water) of *Parreysia favidens***

| Sl.No. | Weight of the Ovary (In mg) | Mean Count of Ova (Y) (in 0.1) | Y x $10^4$ (in 1 ml) | Total Count in 3 ml | Standard Deviation | Standard Error |
|---|---|---|---|---|---|---|
| 1 | 100 | 109-6666 | 1096666 | 3290000 | 13.57699 | 9.60035 |
| 2 | 180 | 109-0000 | 1000000 | 3000000 | 13.89244 | 9.82644 |
| 3 | 175 | 111-3333 | 1113333 | 3340000 | 12.8582 | 9.09212 |
| 4 | 120 | 177-6666 | 1776666 | 5330000 | 42.35957 | 29.95274 |
| 5 | 100 | 115-3333 | 1153333 | 3460000 | 19.29594 | 13.64429 |
| 6 | 99 | 99-6667 | 996667 | 2990000 | 14.74222 | 10.42433 |
| Grand Total | | 18944 | 3568333 | 294478.1 | 131694.6 | |

According to Bloomer (1931) the follicles show the presence of globules in *Lamellidens marginalis,* Patil and Bal (1967) observed the seasonal gonadal changes in adult mussels. The spawning starts in the month of march and continues upto October indicating a prolonged breeding period. After the starting of spawning, lipid globules appear in the lumen of the follicules. In the male, spermatic merulac appeared in male individuals during the breeding season, no indeterminate sex condition or hermaphroditism was noticed. The sexually inactive stage (resting stage) was very short. In the present work the reproductive capacity and fecundity of *Parreysia favidens* were determined. Patil and Bal (1976) have found seasonal variation of chemical components of ovary of *Parreysia favidens*. Proteins, glycogen and lipid and inorganic constituents vary in different seasons. These variations were correlated with the gonadal changes of the mussel.

Agrawal (1980) observed in the gonads of *Indomala caerulae* (Lea) partial spawning in the month of March that gradually rose till July and August. The species is considered to be a continuous breeder throughout the year, having a peak of spawning in the months of October and November. After the onset of spawning lipid globules appear in the lumen of the follicles. The spermatocytic morulae appeared after the start of the spawning.

Bauer (1987) postulated reproductive strategy of long extended period of reproduction in the freshwater pearl mussel, *Margaritifera margaritifera*. Due to the extended reproductive period populations are less vulnerable to fluctuations of environmental condition. The high fertility was independent of age and absence of a postreproductive period, resulting in a single female producing $200 \times 10^6$ glochidia during its reproductive life span of 75 years.

The breeding sesson of *P. favidens* is similar to the findings of Nagabhushanam and Bidarkar (1990).

The breeding season in *P. favidens* falls between April to July as shown by size frequency study of ovaries. It may be concluded that the months of April to July comprise the breeding season of mussel, *P. favidens*. The spawning takes place during the last week of May and is continued till first week of July.

The reproductive cycle of a freshwater bivalve *Parreysia favidens* (benson) of Kosi river has been investigated by determining the frequency percentage of different size ova present in the ovary in different months of the year.

In December the ova were of very small size (0-0.75 μm). They increased in size from February onwards. Three size groups of ova were encountered of which 15 per cent were in the size range of 0.75-1.50 μm. In April four size groups of ova were encountered of which 6.7 per cent ova were in the range of 2.25-3.00 μm. In May seven size groups of ova were found of which 13 per cent were in the size range of 2.25-3.00 μm and 19.5 per cent were in the size range of 3.00-3.75 μm. The largest sized ova were I the range of 4.50-5.25 μm. The small size groups of ova were in the range of 0.0-0.75 μm.

An individual animal with body weight 30.1g and having ovary mass of 120 mg produces 53,30,000 (Fifty three Lakh and thirty thousand) ova.

April to July is the breeding season of *P. favidens*. In the last week of July the mature ova are discharged. During this period the gills become thickened, thousands of eggs are discharged in the mantle cavity of the body and are carried through the ostia into water-tubes of the outer and inner gill laminae. The fertilized eggs are hold by mucus secreted by gills which develop into the Glochidium larvae.

# 6
# Thermal Springs

## Introduction

Thermal springs are high temperature aquatic ecosystem. These water bodies are of special interest from ecological and evolutionary point of view. In such biologically simple environments experimental microbial ecology is most easy to carry out, since microbial interactions are minimized or absent. A thermal spring usually maintains a constant temperature and flow of water and thus it is a "Laboratory in the natural condition'. No man made thermostat is as precise as to keep temperature constant for years together. A thermal stream as an ecosystem is of very simple type, having relatively a few species and short food chain helps in studying productivity, trophocynamics, population fluctuation and species interaction with very little variation.

Not all temperatures are equally suitable for the growth and reproduction of the living organisms and it is therefore essential to consider which thermal environments are most fit for living organisms. For such a study, high temperature environments are of special interest in that they reveal the extremes to which evolution has been pushed. The high temperature environments most useful for study are those associated with volcanic activity, such as hot springs, since these natural habitats probably existed throughout most of the time in which organisms have been evolving on earth. Unfortunately, the geochemist has different interest than biologist, and therefore assays different substances. For the geochemist, substances like silica, potassium, sodium and rare gases are of interest, whereas biologist is interested in nitrogen, phosphorus, organic carbon, sulphide which are essential for the growth of the organisms.

Thermal springs are widely distributed throughout the world but are most numerous in the areas that are volcanically active primarily during tertiary. The early geochemical work is reviewed by Allen and Day (1935). chemical analysis have also been made by biologists seeking an explanation of the reputed curative properties of certain springs (Waring, 1965). Many hot springs have significant amounts of hydrogen sulphide. The concentration of such interesting elements as fluoride, arsenic, rare earths and gold varies very much from spring to spring. Many springs are highly radioactive whereas others have no more radioactivity than normal ground waters. Some springs precipitate silica, others deposit travertine ($CaCO_3$) and still others form elemental sulfur.

The water of many thermal springs are either neutral or alkaline consisting essentially of dilute sodium chloride-bicarbonate solutions with small amounts of trace elements and biologically important anions. The pH of these springs is controlled by a bicarbonate-buffer system and varies with the period pressure of $CO_2$ gas. Another major group of hot springs consists of dilute solution of sulphuric acid with low pH values (2-4). These groups are often high in metal ions, ammonium ion and phosphate. A few hot springs are mixed chloride sulphur waters, usually with pH between 4 to 6. Another interesting group of springs called travertine springs, are rich in $CO_2$ and super saturated with $CaCO_3$ and produce extensive deposites of travertine as $CO_2$ diffuses out of the system.

In India thermal springs are scattered throughout the country and occur either as solitary or in groups, at different altitudes ranging from sea level to nearly three thousands meter above the sea level. Early investigations in India were mainly on the medicinal properties of the springs (Newhold, 1848; Buist, 1851; Mac-pherson, 1855; Schlagintweit, 1862). Oldham and Oldham (1882) realizing the importance of thermal springs, prepared an exhaustive catalogue on Indian hot springs. They reported more than three hundred thermal springs scattered in different parts of the country.

The radioactivity and therapeutic value of several springs have been reported by different workers in the early part of the twentieth century. The estimation of some thermal springs of Bombay-Sind-Baluchistan were reported by Sierp and Steichen 1911), Stecikchen 1912 and Sierp 1913. Some springs in Bombay have been recorded as being permanently radioactive. This is due to the presence of radio emanation or radon in solution, the radon being derived from the disintegration of radio-active elements present in the rocks through which the water circulates. Since radon has a half life of only 3.825 days, water of such springs completely lose their radioactivity stored for a long time. Nag 1931, and Chatterjee 1936 studied the radioactive properties of Rajgir thermal springs. Ghosh 1948 investigated the radioactivity in India. Chatterjee and Chatterjee 1958 investigated the radioactivity of various thermal springs at Bakreswar which ranges from feeble to very strong radioactivity. They recorded Agni Kund to be very strongly radioactive (7.3 µm Mc/1) and Swetaganga as feebly radioactive.

LeTouche (1918) prepared a list of springs reputed to have therapeutical properties. Ray (1932) extensively studied the mineral waters of India dealing with medicinal properties. Ghosh 1948 emphasised the importance of geological and chemical analysis, Classification and radioactive properties of varios thermal springs of India. He studied in all 112 springs including the important springs of Bihar. He recognized 4 main broad belts (Figure 6.1) in which majority of the thermal springs occur.

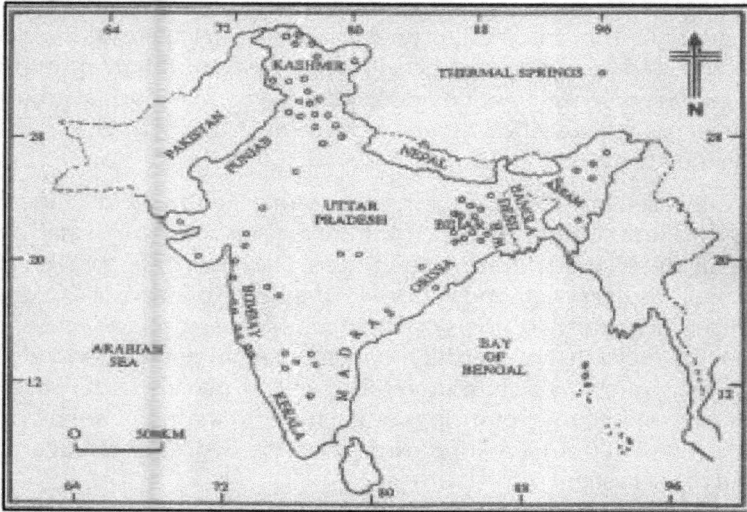

**Figure 6.1: Distribution of Thermal Springs in India.**

1. Bihar
2. Along the West Coast of India
3. Sind-Baluchistan area and
4. The Himalayan belt.

So far as their chemical nature is concerned they have been classified into four main groups:

1. Simple or Indifferent waters, having low mineral contents;
2. Alkaline waters
3. Sulphur waters and
4. Chloride or saline waters.

Kirtikar (1886) was the first to report a thermal alga, *Conferva thermalis* Birdwoodi in India. Since the middle of the twentieth century thermal algae attracted the attention of many biologists. Drouet (1938) reported twelve species of blue green algae from the hot springs of Ladak. Since then Gonzalves (1965) and Vasistha (1968) investigated the thermal algae of India. These workers mainly studied the systematices and upper temperature limit of life.

In recent years a number of studies concerning primary productivity, nutrient cycling, tropho-dynamics and succession of communities have been undertaken in hot spring which will help in understanding the hot spring ecosystem. Brock (1967) a,b; 1970 and 1973; Brock and Brock (1966,19671,b); (1968a,1969a,b) and Castenholz (1967 a,b); (1970 and 1976) exhaustively studied the different aspects of thermal spring ecosystem and reviewed extensively the environmental requirements and biology of thermophilic algae. Stockner 1968a and Naiman 1976 investigated the primary productivity in thermal stream. Kulberg 1968 and 1971 investigated the greeing of blue green-algae, grazing flies and water mies in a thermal effluent of Yellowstone Park.

In India, no comprehensive study has yet been done on diel cycle of abiotic factors, analysis of periphyton, biomass estimation of algal mat, seasonal variation of biota in relation to abiotic factors along the hot spring thermal gradient. Moreover, the thermal springs located in Munger district (Bihar) are very much neglected. There are seven groups of thermal springs in Munger district and with the exception of Sitakund-Phillips Kund group at the northern extremely, all the thermal springs are located in the Kharagpur hills (Figure 6.2).

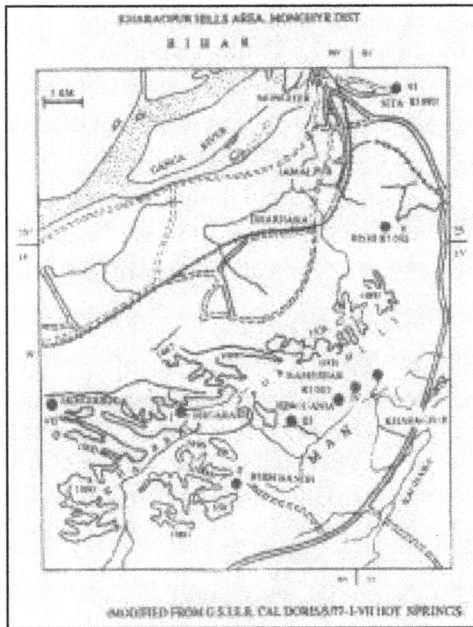

**Figure 6.2: Location Map of Hot Springs.**

These are:

1. Bharari (Choraman) Group;
2. Bhimbandh group;
3. Hingania group
4. Rameshwar – Lachmishwar-Bhowrah Kund group;
5. Rishikund

6. Sitakunbd-Philips Kund group and

7. Sringirikhi group;

All these springs seem to lie in a belt, being Sitakund Phillips Kund group at the north. Then the area in which other springs are located take a southerly direction to the Rishikund, Rameshwar-Lachmishwar-Bhowrah Kund and then turn to the south west to the springs of Bhimbandh (25°3′ N; 86°25′E) and Bharari (25°7′ N; 86°21′E). The spring waters of Bhimbandh have the temperature range of 35-63°C, with an estimated total flow from all the springs of Kharagpur hills being 5 to 5.3 Cu m/sec. these waters have low mineral content and acidic pH val;ues except Sita-kund Phillips Kund group whose pH varies between 6.0 to 7.5 (Saha and Datta Munshi, 1983).

Macpherson (1855) pointed out the possibilities of converting the thermal springs of India into spas and sanatoria. There is enough scope for the development of these thermal springs into spas which will not only attract people of different parts of our country but also foreign tourists. It is generally believed that the thermal spring waters have many therapeutic properties. The spring water which contain radon can cure skin disease, gout, rheumatism and induce apertite and correct many types of metabolic disorders.

According to Brues (1928) the physical environment of a thermal spring differs from that of typical freshwater pond or stream in the following important respects:-

i) An abnormally high and constant temperature,

ii) The presence of certain Salts in considerable amounts,

iii) Deficiency of dissolved oxygen and excess of $CO_2$ and $H_2S$.

All these characters seem to be important in determining the composition and distribution of biota in hot springs.

Brock 1967a pointed out that the hot springs form an ideal ecosystem for the study of many ecological problem, since they provide constant environment. He (1970) pointed out that the species diversity is low in the hot springs and often only a single species may be present. The relative constancy of the thermal stream environment affords an excellent opportunity to estimate the seasonal changes in algal standing stock, growth and production.

## Origin and Geology of Thermal Springs

This term "Thermal spring" has been subjected to variable meanings. Gillbert (1875) listed as thermal only those springs which exceeded the mean annual temperature of the air by 15°F. According to Meinzer (1923) thermal springs are those having a temperature significantly above the mean annual temperature of vicinity. He further sub-divided them into hot and warm springs,the former having higher and the latterlower temperatures than the human body. The water of springs having temperature in the range of 20-30°C is not considered thermal in India whereas in Iceland, spring water of the same temperature would be thermal.Therefore, from geological point of view the thermal springs should be defined relative to the mean annual temperature because in India, any ordinary spring water will emerge at a higher temperature than water from an ordinary spring in Iceland.

Investigations have been conducted on thermal spring areas with a view to elucidate their geological mode of occurrence. Thermal or hot springs are found scattered all over the world and are especially abundant in volcanically active regions such as Japan, New Zealand and Iceland. The Thermal springs generally occur in the following four geologically characteristic areas:

a)  along faults, fissures, shear zones

b)  along the nose of folds

c)  along the contact plane of two adjacent rock types, and

d)  along solution cavities.

## Sources of Thermal Water

The constancy of temperature of the water of thermal springs indicate that it comes out from considerable depth below the surface. Two different types of sources of thermal water have been proposed.

### a) Vadose or Geothermic Water

Meteoric water from the surface descends into the interior through some extended fissures or along the surface of an impermeable layer. It goes below the ground water level and is made to circulate there under the combined action of gravity and capillary forces. As the temperature of the earth crust increases with depth roughly at the rate of $1°C$ per 31.7m, the water becomes heated as it descends. When it is driven upwards by the disposition of the geological structure, the water emerges as a thermal spring.

But this explanation is insufficient for those thermal springs having high temperatures and mineralization.

### b) Juvenile or Hypogenic Water

Suess (1902) hypothesized that the juvenile water is formed deep inside the lithosphere by synthesis from oxygen and hydrogen, since at great depth oxygen is fixed in the volcanic magmas. Suess (1902) proposed that the synthesis results at high temperatures and under great pressure by the combination of juvenile hydrogen with atmospheric oxygen which has penetrated long ago into the crust of the earth.

The constant temperature of the thermal springs is maintained by the mixing of two waters, magmatic and meteoric water, at some depth below the ground water level where the quantity of the surface water is fairly constant being dependent on the pore space, cracks and crevices of the rock bed and the mean annual rain fall but independent of the fluctuation of precipitation.

Suess (1902) believed that volcanic hot spring water is dominantly magmatic in origin. Allen and Day (1935), from their extensive studies on the thermal spring areas of Yellowstone National Park, U.S.A. concluded that the thermal springs are essentially circulating ground water of surface origin, which got heated and augmented by steam originally in a superheated state rising from an underlying magma through deep cracks in the earth's crust. Isotopic studies of thermal springs revealed that the water is mainly meteoric in origin with little fluids added from some other sources (Craig et.al. 1956).

Generally, speaking thermal springs are outlets for large hydro-thermal circulation systems. These systems consist of (1) area of recharge and downward percolation (ii) zones of subsurface flow and heating and (iii) zones of ascent and discharge. A single hydro-thermal system may include several regions of outflow, each of which constitutes a separate thermal region (Bodvarson, 1966).

Hydrothermal systems in zones of 'Recent Volcanism' are classified as high temperature areas and in the region of 'Tertiary Volcanism' as low temperature area. In the high temperature areas where the ascending water temperature may exceed 200°C (Bodvarson, 1960), geothermal activity is characterized by steam holes, solfataras and fumaroles but hot water springs are generally absent. On the other hand in low temperature areas, the ascending water temperature usually does not exceed 100°C. Fumaroles and solfataras are absent in these regions and only hot water springs are found.

## Sources of Thermal Energy or Heat

Stearnst *et.al.* (1937) described the following as the cause of the sources of heat in a thermal spring system.

a) As the meteoric water descends down in the interior of the earth, the temperature increases with depth regularly. Such water is driven upwards by thermal expansion pressure exerted by the containing gases and by density difference between descending cold and ascending hot water and is finally discharged to the surface as a thermal spring. It is not unlikely that the water of the hottest springs of Bihar are possibly issuing from depths over 8.16 km.

b) Heat in an underlying magma is transformed to the spring water either through rock conduction or by direct participation of magmatic fluid as in Iceland, New Zealand and Japan. High content of $CO_2$, $H_2S$, Li, B. F. As etc. in such waters is indicative that considerable amount of heat is contributed by the active participation of the magmatic fluid.

c) It has been estimated that 1 cubic km granite can yield about 25.30 millions of metric tons of water from fusion of $H_2$ and $O_2$ which at 1100°C would form $16 \times 10^{16}$ cubic meters of steam and $28 \times 10^6$ cubic meters of other gases. If by fissuring and subsidence in the lithosphere such mass of rock is carried to a depth of 25,000-30,000m it would then be in the heated region, and the formation of vapours under great pressure would occur. Ordinary thermal springs might have been formed by the same process and originated from a sort of distillation of the combined water contained in the depressed mass of rocks. But such a hypothesis the foundering of a large block of the country rock in late geological times and the instability of the country. We cannot, however, detect any geological evidence of such large scale foundering of the earth's crust in the Peninsula since the Gondwana times. Towards north of the tract of thermal springs is the Indo-Gangetic trough, which might have been very deep in the beginning, the lowering of which in the depths of earth may still be taking place.

d) Another view regarding the source of heat seems to be the exothermic reaction of certain chemical and mineral transformation *viz.*, in the disintegration of radioactive elements and the decomposition of pyrite in the generation of $H_2S$ which characterizes many of the springs.

e) Heat is evolved in the zones where faulting of rocks have occurred. Many thermal springs are seen to emerge along active faults and fissures. Some amount of heat in those thermal springs are contributed by the faulting of rocks as in some thermal springs of Western United States.

As most of the thermal springs are associated with volcanic rocks, so many scientists have tended to assume that the source of heat in a thermal spring is due to volcanic activity. Allen (1954) concluded that steam given off by magma is the source of heat in all the thermal springs he studied and thought that the mineral content of the water was partly magmatic in origin.

Geothermal activity is probably responsible for the creation of high temperature environments. The largest concentration of hot springs and fumaroles are found in Yellowstone National Park (USA), Iceland, New Zealand, Japan and the (USSR). Sometimes it is not realized that geothermal hearing is a normal phenomenon throughout earth's surface, although in most areas the heat production is so small that it is unrecognized. Likens and Johnson (1969) calculated the heat production of lakes from geothermal sources, the estimated value of which comes to be $7 \times 10^6$ K. Cal/year. On the other hand, this heat production in a small Fijan hot spring is about $3 \times 10^9$ K.cal/year (Healy, 1960), which is almost three orders of magnitude higher. So, the geothermal flux in the localized region of a hot spring must be very high.

The chemical and few spectroscopical analyses of the waters of the Indian thermal springs indicate the presence of the characteristic components of juvenile waters. The presence of high silica and soda content, is characteristic of the hot springs emerging in Archaen terrain of India. Such composition of the waters of the thermal springs is an evidence of participation of juvenile water in the mineral spring activity. Inspite of such striking similarities the magmatic influence to the hot springs of Bihar is difficult to accept, because there is no outward evidence of late magma activity anywhere in India like. Yellowstone Park (U.S.A.). But observation in bore holes indicate very rapid rate of increase of temperature downwards. So it is concluded that the parent magma which gave rise to the rhyolite imtrusion in Pliocene times, is still maintained in a fluid state in a chamber fairly close to the surface. In case of Bihar Archen terrain, the problem is more difficult. The latest exposition of the Rajmahal volcanics to the north east of the Archaen tract is believed to be Jurassic and there may be a relationship with the Deccan trap. It may be possible that large injections of these traps are still maintained in fluid state below the hot spring zone in Bihar. At present the evidence seems to favour the hypothesis that the source of heat is connected with juvenile waters released by a magma-chamber hidden below the surface (Ghosh, 1948).

## Biota of Thermal Springs

The phenomenon of thermobiosis, including both survival and growth of algae at higher temperature has attracted the attention of naturalists for about last two centuries. At the end of the eighteenth century, Saussure (1776) reported both plants and animals in some European hot springs. Since then it has become more and more apparent that the biota of hot springs are more resistant to heat than other organisms.

## Flora

The flora of thermal springs is almost entirely algal and is mainly composed of members of Myxophyceae (blue green algae) and bacteria. They represent the most thermal tolerant group of organisms found in nature. As early as 1846, Flourens discovered algae in a hot spring of Iceland at a temperature of 98°C. Hooker(1854) collected *Leptothrix* from the hot springs of Himalayan region at a temperature of 75.5°C. Brewer (1866) found algae in the hot springs of California at 93°C. he also collected some "Nostoc formen" in hot Geyser. Archer (1874), Dyer (1874), Mosley (1874,1879) and Thomson (1877) investigated the nature of algae found in the hot springs of the Furnas valey, Azores. Kirtikar (1886) recorded a thermal alga ( *Conferva thermalis f.* Birdwood) from India. Weed (1888 and 1889), Davis (1897) and Tilden (1897) studied the vegetations of thermal springs of Yellowstone Park. Myxophyceae are of particular interest as they assist in the deposition of large quantities of calcareous travertine and siliceous sinter. West (1902) made phycological investigations of Iceland's hot springs and made taxonomic compilation. Satchell (1903) made extensive studies of algae and bacteria Yellostone Park and reported the upper temperature range for algae to be 75-77°C and for bacteria 89°C. Elenkin (1944) reported the upper limit to be 85°C for the blue green algae. Strom (1921) studied the thermal algae of hot springs of Spitzbergen. The knowledge of Japanese thermal flora was mainly due to Molisch (1926) who discussed the distribution of Myxcophyceae in different hot springs of Japan. Petersen (1928), Dover (1932), Copeland (1936), Schwabe (1936) and Nash (1938) studied extensively the flora of different hot springs. Geitler and Ruttner (1936), in their investigation of the Japanese thermal algae recorded 3 species at 60°C and the largest number of species between 35 and 40. Copeland (1936) gave an excellent and exhaustive account of thermal Myxophyceae of Yellowstone Park. Nash (1938) reviewed the literature on thermal Myxophyceae and studied the algae in relation to certain ecological factors. Mason (1939) reported Oscillatoria at temperatures between 33 and 58°C, and *Denticula* in the temperature ranges of 33-45°C. Tuxen (1944) has made comprehensive study of the Icelandic thermal flora.

Prasad and Srivastava (1965) reported 24 species of Myxophyceae including one new variety (*oscillatoria*) *asorvensis* var. *thermalis*) from four thermal springs (45-97°C) of Himachal Pradesh (India) at altitudes between 1500 m and 2800 m. They reported Aphanothece *saxicola* at a high temperature of 83°C. In the same year Thomas and Gonzalves (1965) extensively studied the algal flora of twelve groups of hot springs in Western India. Vasistha (1968) studied ther thermal algae of Myxophyceae group from about 116 hot springs having a termperature range of 29-94°C. He recorded 336 species belonging to 58 genera and also studied the physic-chemical nature of the springs.

# Study Area and Biota of Thermal and Cold Streams

## Study Area

### *Climatology*

Bhimbandh area (Munger) falls in the subtropical belt and as such the climate is characterised by high summer temperature, humidity and well distributed precipitation during monsoon. There are three distinct seasons in a year in this locality:

 i) Summer (March to June)

 ii) Rainy or monsoon (July to October)

 iii) Winter (November to February)

The average monthly temperature, rainfall and relative humidity are shown in Figure 6.3. The summer begins from about March and lasts upto June. The low pressure zone develops over north west India due to which a sea wind arises which is not strong enough to start monsoon. But it brings humid air over the continent which explain relatively higher relative humidity in North Eastern India including Munger during the summer. Wind velocity is higher especially in May and June (9-12 km/H) when hot dusty winds locality known as "LOO" blow throughout the day. The average maximum temperature ranges between 35.6 and 37.1°C.

In June due to shifting of low pressure zone towards North East India by the showing of polar circulation in the north, strong monsoon brings torrential rains. The South westerly winds are strengthened by an extension of the hemisphere trade winds. The winds entering the bay of Bengal are deflected to the north and then to north-east by the Himalayas which explain the direction of monsoon from the South east at Bhimbandh, Munger. The weather remains generally cloudly and sultry. Heavy precipitation generally occurs in July and August months (206-352mm). The average rain fall for the last twenty years is about 102cm and 75 per cent of rain fall was recorded during monsoon. The relative humidity always remains more than 80 per cent. The average maximum temperature ranges between 30.8 and 35.4°C.

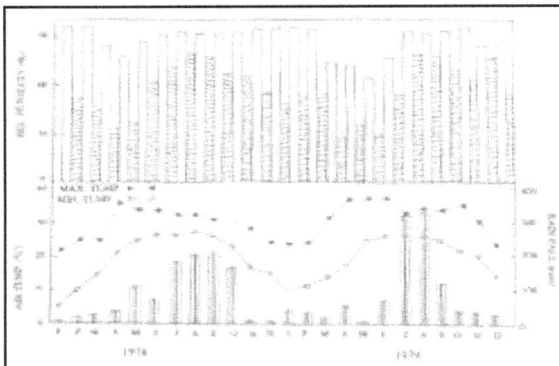

**Figure 6.3: Climatic Conditions of Study Area**

The winter extends from the middle of November and lasts upto the end of February, India is projected from the effects of the Siberian anticyclone by the Himalayas. Generally, westerly wind passes over the Gangetic plain in which Bihar (Bhimbandh) is located. This cold current leads to a distinct fall in atmospheric temperature. This current passes eastwards and meets the north east trade wind and flows over the Gangetic plain carrying some moisture that leads to some precipitation during winter. The rain fall is however, less than 85mm at Bhimbandh in winter. January is the coldest month of the year (4.3-9.7°C).

The average monthly temperature, rainfall, relative humidity with velocity and wind direction are presented in (Figure 6.3).

## Physiography

The thermal springs under investigation are located in Bhimbandh sanctuary in the Kharagpur hills in Munger District, Bihar. The entire area of the sanctuary (68190.172 Hectares) is compact and contiguous in one hill block. The sanctuary tract lies between 24°55 to 25°15' North latitude and 86°15' to 86°33' East longitude. The entire sanctuary area is owned by the State Government and legally constituted "Protected Forests" under the Indian Forest Act (Act XVI of 1972). The sanctuary is surrounded by railway tract passing from Kiul to Sahibganj loop section of the Eastern Railway on the north and Howrah-Kiul main line section on the west. The state highway passes from Jamui to Munger on its southern and eastern site.

The sanctuary is full of wild life. It was the famous hunting ground of Banaili Raj. The sanctuary still abound in tiger, panther, sloth bear, sambhar, barking deer, wild boar, nil gai, antelope etc. The entire sanctuary is covered by deciduous forest. The flora mainly consists of *Shorea robusta, Terminatia somentosa. Buchanamia latifolia. Terminalia arjuna, Croton oblongifolias, Diospyros melanoxylon, Smilax prolifera, Bawhinia vahli, Milletis suriculata, Acacia pennata etc.*

Bhimbandth group of thermal springs (latitude 25°3' North longitude 86°25 East; (90m above sea level) are located about 1 km north west of Bhimbandh village and is about 90 km from Bhagalpur and 63 km from Munger town. It is only 10 km away from the Jamu-Munger highway. There are two main location of hot springs at Bhimbandh (i) Chhockipani, near the Bhimbandh rest house (ii) tatalpani, about 100m further to the north, at the foot of the horn stone hill "Damadama" hot water spouting from the ground in every direction. The principal springs, 8 to 10 in number arise within a space of about 250 sq. meter. The waters of all these hot stream forming a rivulet called the "man" (Sherwill, 1832). Not far off is a small pool of cold water under an overchanging rock called Bhimkund which is named after Bhim of Mahabharat fame and is visited by pilgrims. From this pool a cold stream originates which finally meets with the "Man". Temperature of the hot springs of Bhimbandh varies from 54 to 63°C. The total estimated flow is about 0.84 to 1.20 cu.m/sec. Copious gas bubbles are seen. Radon content varies from 0.7654 to 1.224 m Mc/litre.

It is a fact that from time immemorial the inhabitants of the country have been aware of the values of mineral springs. Thermal springs which occur in inaccessible places become generally the places of pilgrimage and annual fair, the water being used for drinking and hot bath. It is, therefore, not surprising to discover that several sacred places of pilgrimage in India are at the sites of well known thermal springs. An excellent historical and mythological account of the hot springs of Munger have been given in District Gazetteer of Munger Roy Choudhury, 1960). It is interesting to find historic testimony to the former existence in this region of an active volcano, as reported by the Chinese pilgrim,. Hieun Tsiang, who visited the neighbourhood of Munger in the first half of the 7th century A.D. he has recorded that by the side of the capital and bordering on the Ganges river is no mountain from which are an land belched forth masses of smoke and vapour, which obscure the light of the sun and moon. Roy Choudhury (1960) mentioned that at Bhimbandh, the high temperature recorded by Dr. Buchaman on 21st March, 1811 was 150°F (65.6°C). Sherwill in September, 1847, Waddell in January of 1890 and Schulten in August of 1913, observed temperature of 147°F (63.4°C) and 148°F (64.4°C) respectively. Accordingly to Mr. V.H. Jackson, there are twelve sources in the Mahadeva group and at least nineteen in the Damadama group and hottest of them may not have been observed. Readings taken between 1912 and 1919 varies from 145.5 to 146°F in the Mahadeva and from 148 to 148°F (64.4°C) respectively. Accordingly to Mr. V.H. Jackson, there are twelve sources in the Mahadeva group and at least nineteen in the Damadama group and hottest of them may not have been observed. Readings taken between 1912 and 1919 varies from 145.5 to 146°F in the Mahadeva and from 148.8°F in the Damadama series. According to Waddell, nearly all the hot springs of Munger are worshipped by the Hindu and semi aboriginal villagers in the vicinity of these outbrasts of heated water. The phenomena of developing clouds of vapour from boiling water of hot streams are regarded by the natives as super natural and as the special expression of the presence of deity. The deity usually worshopped at the springs by the natives and aboriginals is Mata or Mai, the mother Goddes, one of the form of Kali, and large melas (fairs) are held in her honour. The more Hundred people however, believed that their favourite God Mahadeva is specially present at all these hot springs and to Him they offer worship.

Kharagpur hill is composed of Archaon quartizine alterating with Agrillites of Dharbar age, in tightly folded inclines. The hot springs occur in fault zone in Archaon quartizing striking NNE-SSW. The age of fault is believed so be Miocene to Pleistocene corresponding to the late phase of movement of Himalayan orology. Geologically thus, Kharagpur hills are composed of two different type of formations broadly the eastern part of the hill is composed of lower Precambrian while the western portion consists of metamorphic sediments of Dharbar age. The hot spring of Bhimbandh issue along the upturned limb of quartizine (NE-SW/SE) along an apparent fault line.

## Morphometry of the Thermal Stream

A number of thermal springs emerge through the fissures from the different places of the foot hills called tatalpani or Taptapani (*i.e.* Hot water). One of the sources recorded the highest temperature (63°C), named as source (A) of the present study. The temperature of other souces varied between 56-60°C. The water from these sources run down, flows northerly through a deep forest, coverage to form a small stream.

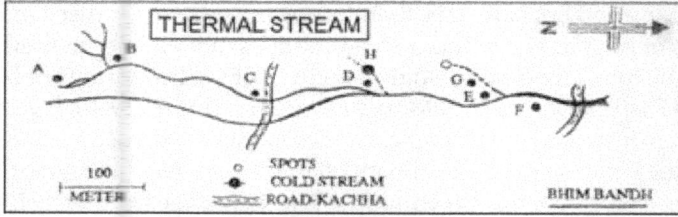

**F gure 6.4: Thermal Stream of Bhimbandh.**

Five more sampling stations (B,C,D,E and F) were selected at 40 to 160 m intervals along with the thermal gradient of the hot stream on the basis of temperature of water. This hot stream meets two cold streams, one (g) at a distance of 500 m and the other (h) at a distance of 350m away from the source. A sampling station, (G) was selected along the cold stream (g), was also included in this investigation. All these sampling stations (Figure 6.4) are situated in the deep forest, shaded by trees except the source (A). Finally the water of these springs form a small rivulet called the 'Man'.

**Sampling Station (A)**

This is one of the source of thermal springs of "Tatalpani" group. It recorded the highest temperature (63°C). It is located at the foot of the horn stone hill "damadama". Water discharges through a number of fissures and is stored in a shallow basin of 1.2m in diameter and 3-10 cm in depth. The water flows north-westerly and meets with other hot streams at about 30 m away from station (A) so form a broad stream.

**Sampling Station (B)**

It is situated about 40 m away from station (A) on the north western side of the hot stream. Here the stream is much broader, ranges between 3.5-4 m but 10-12 cm in depth, only and shaded by trees. The temperature varied between 50-57.5°C. Water is transparent throughout the year.

## Biota of Thermal Stream

The biological characteristics of thermal water depend upon several factors of which temperature seems to be the most potent one. Life exists mostly in the form of flora at higher temperature gradient. Animal life appears only at lower range of temperatures.

## Flora

The flora of hot stream along thermal gradient of Bhimbandh mainly comprises a stage, occurring both as phytoplankton and periphyton. They play a key role in the productivity of the thermal stream. The algal flora is mainly represented by Myxophyceae or Cyanophyceae (Blue Green algae), Bacillariophyceae (Diatoms) and Chlorophyceae (Green algae) as shown in (Table 6.2.)

## (A) Myxophyceae

The Myxophyceae is a distinctive group of algae in which the pigments are localized in the peripheral portion of protoplasm and include chlorophyll *a, carotene,* distinctive xanthophyll,blue pigment (C-phycocyamine) and a red pigment (C-phycoerythrin). Another unique feature of Myxophyceae in the primitive type of nucleues, which lacks nucleolus and a nuclear membrane.

### Table 6.2: Algal Flora of the Hot Stream of Bhimbandh

| Sl.No. | | Plankton | Periphyton |
|---|---|---|---|
| | **MYXOPHYCEAE** | | |
| 1. | *Synechococcus lividus* | 63.32 | 63-38 |
| 2. | *Synechocyatis sallensis* | 57.5-46 | 57.5-46 |
| 3. | *Aphanocapsa brunnea* | 50.5-39 | – |
| 4. | *Chroococcus minutes* | 54.5-32 | 54.5-44.5 |
| 5. | *Aphanothece* sp. | – | – |
| 6. | *Phormidium atricanum* | 58-32 | 58-38 |
| 7. | *Oscillatoria limosa* | 50-42.5 | 50-32 |
| 8. | *O. proboscidea* | 40-32 | 50-32 |
| 9. | *O. jasorvensis* | 46-39.5 | 50-32 |
| 10. | *O. chlorine* | 46-45 | 50-32 |
| 11. | *Lynghya martensiana* | 48-32 | – |
| 12. | *Spinulima subsalsa* | 48-32 | – |
| 13. | *Mastigocladus laminosus* | – | 49.5-47.5 |
| | **BACHLARIPHYCEAE** | | |
| 1. | *Pinmularia viridis* | | 44-41 |
| 2. | *P. interupta* | – | 57-39.5 |
| 3. | *Navicula peregrine* | 49-39 | 57-32 |
| 4. | *Cymbella cistula* | 49-39 | 47-39.5 |
| 5. | *C. turgida* | – | 47-39.5 |
| 6. | *Gomphonema sphacrophorum* | – | 47 – 38 |
| 7. | *G. lanceolatum* | 47.38 | – |
| 8. | *Diploneis subovalis* | 43-39 | 43-42 |
| 9. | *Fragilaria brevistriata* | – | 57-39 |
| 10. | *F. intermedia* | 47.5-32 | 45-38 |
| 11. | *Synedra ulna* | 44.5-39 | 43-38 |
| 12. | *Eurocia monodon* | 40.5-39 | 57-32 |
| 13. | *E. pectinalis* | 49-39 | 57-38 |
| 14. | *Suriwella tenera* | 47.39 | 44.5-39 |
| 15. | *S. elegans* | 40.5 – 39 | 44.5-39 |
| 16. | *S. robusta* | – | 44.5-39 |

*Contd...*

**Table 6.2 –*Contd...***

| Sl.No. | | Plankton | Periphyton |
|---|---|---|---|
| | **CHLOROPHYCEAE** | | |
| 1. | *Ocdogonium* sp. | 47.5-45 | 45-39 |
| 2. | *Spirogyra singularis* | – | 43-40 |
| 3. | *Netrium digitus* | – | 44.5-40 |
| 4. | *Closterium ramidum* | 40.5-38 | 44.5-38 |
| 5. | *C. setaceum* | 45-38 | 40.5-38 |
| 6. | *Cosmarium amoenum* | 47-38 | 45-38 |
| 7. | *C. tenue* | – | 44-38 |
| 8. | *C. punctualatum* | – | 45-38 |
| 9. | *Staurastrum surgeacens* | – | 42-40.5 |
| 10. | *Euastrum subloratum* | – | 42-40.5 |
| 11. | *Pleurotaenium* sp | – | 40-39 |

These algae can tolerate very high range of temperatures and form the dominant group of thermal springs biota. Altogether 10 genera representing 13 species of Myxophyceae, were collected (Figure 6.5). These are:

**Order: Chroococcales Wettstein**

**Family: Chroococcaceae Nageli**

*I. Synechococcus lividus Copeland* (**Figure 6.5-1**)

Cells cylindrical usually straight or sometimes feebly curved, 1.8-2.2µ in diameter and 9-19.5µ long; separating soon after division, but frequently in linear pairs. Cell contents usually with one or two granules, polar in position Olive green to dull blue green.

This was the only algal flora found as such high temperature as 63°C forming dense algal mat in association with *Pharmidium africanum* in the temperature ranges of 50-55°C.

## (B) Bacillariophyceae

The Bacillariophyceae include a large number of unicellular and colonial genera which differ from other algae in the shape of their cell. The main characteristic feature of diatoms is the presence of highly silicified cell wall which is composed of two overlapping halves (valves). Only Pennales type of diatoms were found in the thermal stream of Bhimbandh. They were collected from 56°C but were abundant in the lower temperature range in the down stream. Altogether 9 genera representing 16 species have been encountered during the study period (Figure 6.6). These are:

**Family – Naviculaceae**

**1. *Pinnularia viridis* (Nitzsch) Her**

Valves linear with slightly convex margins and rounded ends. Axial area narrow,

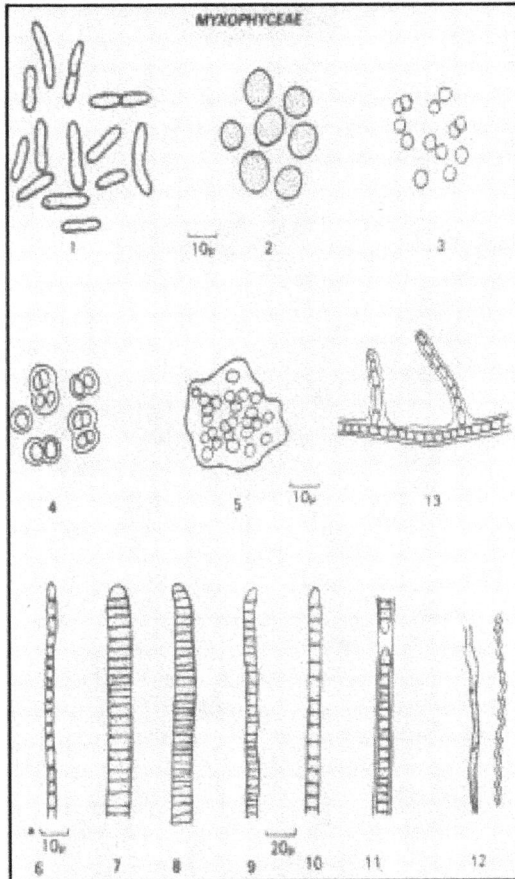

**Figure 6.5: Myxcophyceae of Thermal Stream.**

slightly widened in the middle. Striae coarse, slightly radial in the middle and convergent at the ends. The longitudinal band is present.

Length: 96-125 µ: Breadth: 19-21 µ

## (C) Chlorophyceae

The Chlorophyceae are a group of algae having their photosynthetic pigments localized in chromatophores which are grass green because of the predominance of chlorophyll-*a* and *b* over the carotene and xanthophylls. Photosynthetic reserves are usually stored as starch. In the thermal stream of Bimbandh, they are only present at the lower range of temperature. Highest temperature recorded for them was 45°C. Altogether 8 genera representing 11 species, were encountered during the study period (Figure 6.7). They are:

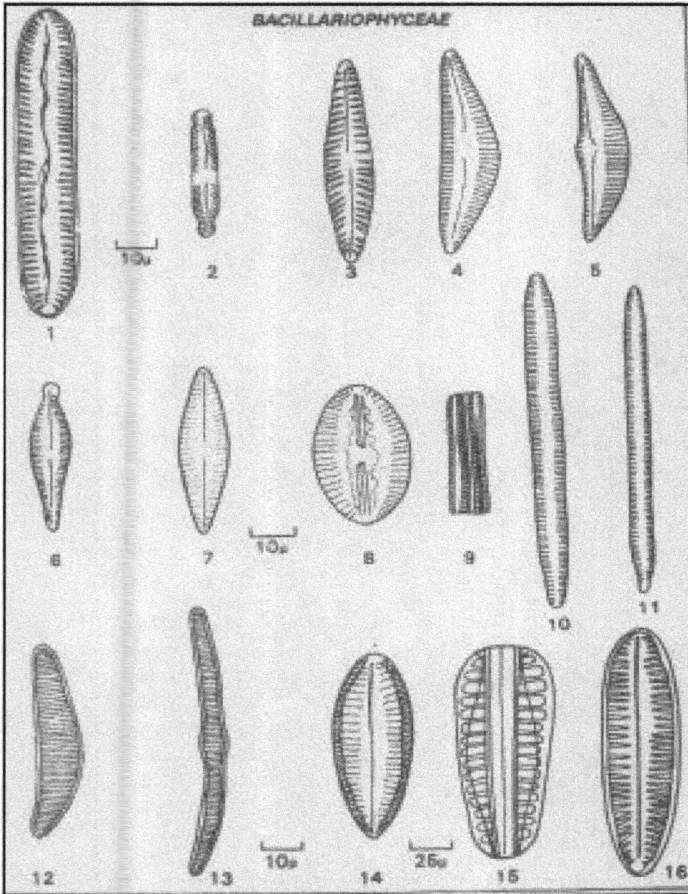

**Figure 6.6: Bacillariophyceae of Thermal Stream.**

### Order: Oedogoniales

### Family – Oedogoniaceae

### Fauna

Fauna of thermal stream is very poor in comparison to flora. The fauna mainly comprises Protozoa, Rotifera, Nematoda, Insecta, Fishes and Amphibia (Table 6.3). Altogether 9 species of Invertebrates and 3 species of vertebrates were encountered during the study period at temperature between 38 and 45°C (Figure 6.8).

These are:

### A) Protozoa

Class – Rhizopoda

Order – Lobasa

Family – Arcellidae

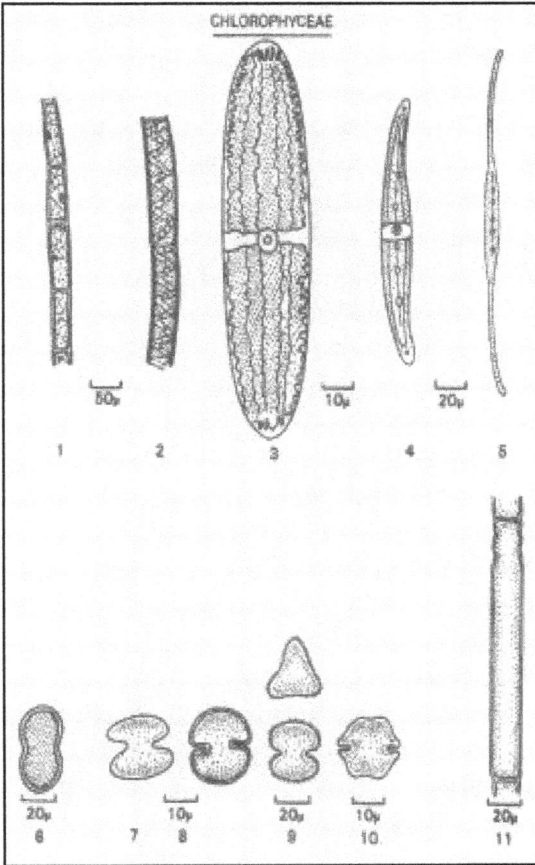

**Figure 6.7: Chlorophyceae of Thermal Stream.**

### 1. *Arcella discoides Ehren* (Figure 6.8.1)

Smooth shell with a large circular aperture, occurred both as plankton and periphyton.

Diameter: 80-125µ

Family – Euglyphidae

### 2. *Euglypha* sp. (Figure 6.8.2)

Shell retort shaped with elliptical scales

Length: 55-75µ; Breadth: 26-35µ

Class – Ciliata

Order – Peritrichia

Family – Vorticellidae

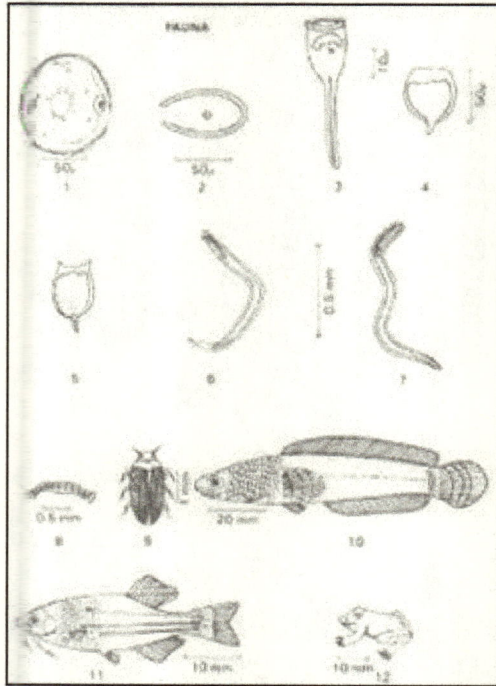

**Figure 6.8: Fauna of Bhimbandh Thermal Stream.**

### *3. Vorticella* sp. (Figure 6.8.3)

Small goblet shaped; about 25µ long with a stalk of 30-45µ in length

### B) Rotifera

Order – Monogononta

Suborder – Ploima

### *4. Monostyla* sp. (Figure 6.8.4)

Common rotifer of freshwater; pseudotroch absent; one toe present

Size: 68 x 92µ

### *5. Brachionus* sp. (Figure 6.8.5)

Trophi malleate, pseudotroch present lorica moderately compressed dorso ventrally;anus is dorsal,situated at the junction of tail and trunk; tail is wrinkled and ends in two slender processes.

Length: 55-76µ; Breadth: 28-35µ

### C) Nematoda

### *6. Monunchus macrostoma Bastian* (Figure 6.8.6)

Two well marked ridges on the pharyngeal wall opposite the dorsal tooth, tail cylindrical; opening of the caudal gland in the middle of the posterior end.

**Table 6.3: Fauna of Hot Stream of Bhimbandh**

| | *Range of Temperature (°C)* |
|---|---|
| **PROTOZOA** | |
| 1. *Arcella discoides* | 46.5 – 40.5 |
| 2. *Euglypha* sp. | 46 – 39 |
| 3. *Vorticella* sp. | 41 – 39 |
| **ROTIFERA** | |
| 1. *Monostyla* sp. | 41 – 38 |
| 2. *Brachionus* sp. | 44 – 42 |
| **NEMATODA** | |
| 1. *Monunchus macrostoma* | 46 – 32 |
| 2. *Trilobus gracilis* | 40.5 – 32 |
| **INSECTA** | |
| 1. *Chironomous larva* | 47 – 37 |
| 2. *Guignotus pradhani* | 43 – 37 |
| **PISCES** | |
| 1. *Channa gachua* | 40°C and below |
| 2. *Danio dangila* | 40°C and below |
| **AMPHIBIA** | |
| 1. *Rana cyanophlyctis* | 40°C and below |

**7. *Trilobus gracilis Bastian* (Figure 6.8.7)**

Mouth is surrounded by 3 pairs of lips, each bearing a well marked papilla.

Length: 1.30 – 1.85mm

**D) Insecta**

Order – Diptera

**8. *Chironomous larva* (Figure 6.8.8)**

The larva usually live in tubes either free or attached to stones

Order – Coleoptera

**9. *Guignotus pradhani nov* (Figure 6.8.9)**

Form elongate attenuated posteriorly and sides almost sub-parallel; dorsal surface pubescent; antennae elongate;

Length: 2.2-2.5 mm; Breadth: 1.12mm

**E) Pisces**

Order – Channiformes

### 10. *Channa gachua* (Ham.) (Figure 6.8.10)

Head 3.5 to 4.2, maxilla reaches to belowhind border of eye, pelvic 2/5 of pectoral length; 4 to 5 scales between orbit and angle of preopercle.

Order – Cypriniformes

### 11. *Danio dangila* (Ham.) (Figure 6.8.11)

Head 5, height of body 3.5 to 4 in total length, lower jaw longer, mouth oblique, maxillary barbells slightly longer, rostral pair shorter than head.Sides with several blue lines; anal fin with 2 or 3 blue stripes.

### F) Amphibia

Order – Anura

### 12. *Rana cyanophlyctis Schncider* (Figure 6.8.12)

Adultsof this species were found on the bank of the hot stream but when approached they dived in the hot stream and tried to hide underneath the rocks. They are also found swimming in the hot stream at temperature ranges of 37-40°C.

# 7
# Lakes and Reservoirs

Lake is a large body of water enclosed by land and occupies a basin. It comes under lentic system and is larger (depth and areawise) than a pond. Lakes are geologically transitory. They are usually born of catastrophes occurring in nature and die quietly (Hutchinson, 1957). Thus a lake lives through youthful stages to maturity, senescence and death when its concave basin is finally filled up.

## Lakes and Reservoir

Besides these natural lakes, man is building up artificial lakes usually called reservoir or impoundments all over the world including areas where there are no natural lakes. Whille not exactly born of catastrophe, man-made lakes are probably also transitory in the geological sense.

Artificial lakes vary according to the region and to nature of the drainage. Generally they are characterized by fluctuating water levels and high turbidity. Production of benthos is often less in impoundments than in natural lakes. The heat budget of impoundments may differ greatly from that of natural lakes depending on the design of the dam. If water is released from the bottom as with the case of dams designed for hydroelectric power generation, cold nutrient rich but oxygen-poor water is exported downstream while warm water is retained in the lake. The impoundment then becomes a heat trap and nutrient exporter in contrast to natural lakes, which discharge from the surface and therefore function as nutrient traps and heat exporters (Figure 7.1).

## Lake Regions

In lakes and ponds three zones are generally evidenced as in (Figure 7.2)

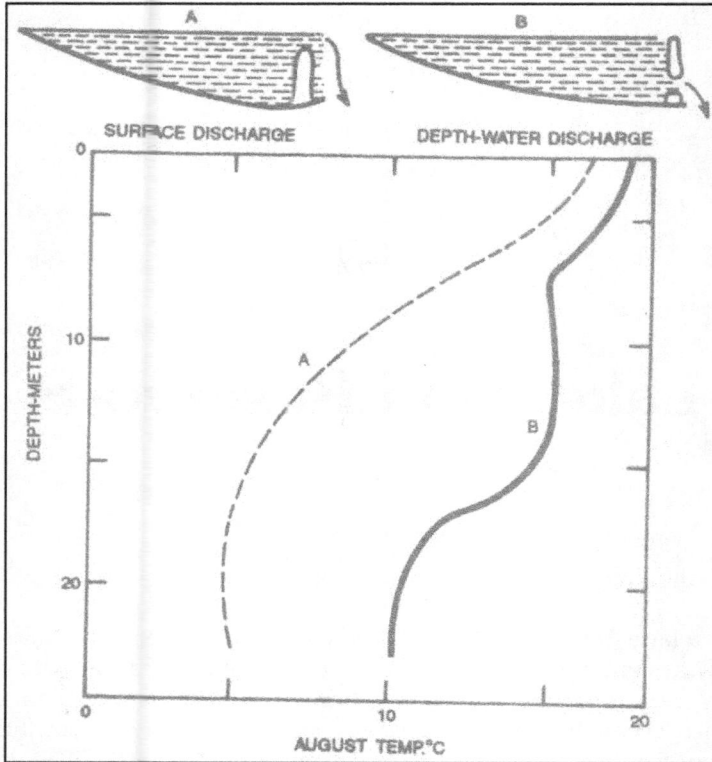

**Figure 7.1: Showing the Process of Discharge.**
**A. Surface discharge in a natural lake. B. Deep water discharge in an impoundment.**

## Littoral (=Litoral) Zone

It is the peripheral shallow-water of lakes and ponds with light penetration upto the bottom and occupied by rooted plants. It is subjected to fluctuating temperature and erosion of shore materials. In certain lakes, sometimes wave action is so much that large aquatic angiosperms are absent and only algae, as benthic mats are present.

## Limnetic Zone

It is the open-water zone (beyond the marginal vegetation) to the depth of effective light penetration, called the compensation level where photosynthesis just balances respiration. In general this zone will have a depth at which light intensity is about 1 per cent of full sun-light intensity. The community in this zone is composed of plankton, nekton (nekton) and sometimes neuston. This zone is absent in small shallow ponds. The term euphotic zone refers to the total illuminated stratum including littoral and limnetic where primary producers fix inorganic carbon to manufacture organic compounds.

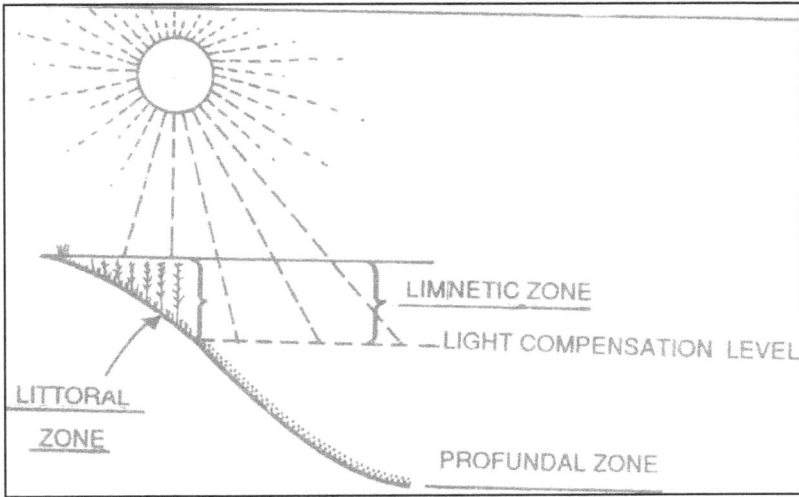

**Figure 7.2: Lake Zones: Littoral, Limnetic and Profundal Zone.**

## Profundal Zone

The bottom and deep water area which is beyond the depth of effective light penetration where respiration and decomposition predominate. It is also called tropholytic region of lake, where breaking down or dissolution of organic compounds or cells take place. The community in this zone composed of mainly animal benthos. This zone is often absent in ponds.

# Origin of Lakes

The major causes for the origin of lakes are given below.

## Glacial Action

Through the action of glaciers lakes are formed (a) either by the deposition of morain and debris *viz.* the Dal and Wular lakes of Kshmir, or (b) by excavation of substratum *viz.* the lakes of Pongong valley in Kashmir.

## Tectonic Phenomena

It refers to the earth crust instability due to warping (twisting), faulting, fracturing, folding, throusting and quaking. There is deformation and adjustment of the earth's shell. The two great ancient lakes of Baikal in Siberia with a depth of more than 1.7 km and Tanganyika of East Africa are formed by tectonic phenomena. Their basins are formed lying between faults and adjacent highlands.

## Land Slides

Many lakes are formed due to the impoundments on stream valleys by rock slides, mud flows, or other mass moverock, example Kumaon lake in Uttar Pradesh.

## Volcanic Phenomena

Volcanoes have modified and shaped past landscape in many areas of the world and continues to do so today. In Japan many lakes are formed by volcanic activities. Tropical countries like Java, Sumatra and Bali have many such lakes. The deepest lake in the United States is the Craster lake of Oregon. This is an extinct volcanic crater, formed when underlying magma (molten rock) has flown out.

## Dissolution of Substratum

Basins are created by dissolution and removal of materials which hold water to become lakes or ponds. Usually the carbonates of calcium and to a lesser extent magnesium are the solutes. Nainital Lake of Uttar Pradesh has been produced by the collapse of surface caused by the removal of limestone by the solvent action of underground water. Many small lakes in the limestone tracts of the Khasi and Jaintia Hills of North eastern region are examples of solution lakes.

## Wind Action

In arid regions, the erosive force of wind removes sand particles from some place and thus lake basins may be formed in the depressions. Further deposition of these removed particle from one place may block the existing stream of other place giving rise to a lake. The salt lakes of Rajasthan (like Sambhar lake) come under this category.

## River Actions

Rivers create standing water habitats of various types. Flood plain lakes are examples which are again categorized as Levee lakes and ox-bow lakes. Levee lakes are shallow often elongate bodies of water that lie parallel to the river bed and are separated from it by strips of higherland. The levee is a natural embankment built up by river sediments deposited at times of high water. The second type of flood plain lakes is the ox-bow lake formed from isolated loops of meandering rivers. These crescent shaped basins are usually deeper than the levee lakes. In the lower reaches of the Ganga and Brahmaputra rivers many such levee and ox-bow lakes are present.

## Biotic Action

Many North European lakes occupy depression in peaty deposits. Dams of sphagnum and other bog plants have been reported by Hutchinson (1957). Beaver damming and man's similar activity are well known. In both instances streams are impounded by putting stream side vegetation or bamboos to trap fish. With the passage of time, reservoirs or man-made tanks are constructed for multiple objectives, namely for irrigation, hydroelectric power generation, flood control, public supplies, recreation, development of fisheries etc.

In principle reservoirs are created by constructing stone masonary or concrete bundhs across a stream. These have penstocks, irrigation release tunnels and spillway as safety mmeasures against floods. Important reservoirs of India are Maithon (Bihar and West Bengal), Hirakund (Orissa), Gandhinagar (M.P.), Ukai (Gujarat), Nagarjunasagar (A.P.), Stanley (Tamil Nadu), Tunga-Bhadra (Mysore), Bhakra Nagal (Punjab) and Rihand (U.P.).

# Lake Characteristics

## Thermal Stratification and Circulation in Temperate Lakes

In temperate lakes, during summer the top waters become warm, than the bottom waters and as a result only the warm top layer circulates. It does not mix with the more viscous cold water, creating a zone with steep temperature gradient in between called the thermocline. The upper, warm circulating water is the epilimnion ("surface lake") and the colder, noncirculating water is the hypolimnion under lake (Figure 7.3). The magnitude and direction of the water is the hypolimnion under lake (Figure 7.3). The magnitude and direction of the water movement depends upon the velocity and duration of the wind. Steady vigorous shore winds may set up return currents which extend to the opposite side of the lake. Encountering the colder denser water of the thermocline region, this down flowing returning current is diverted in a horizontal direction and flows toward the opposite shore, maintaining a level above the thermocline. In such a circulation, the upper stratum of the epilimnion flows toward one shore, the lower-stream (the return current flows toward the opposite shore and the stratum which is known as the shearing plane is practically without motion. The name of the region in between epilimnion and hypoliminion was described as thermocline by Bronsted and Wesenberg-Lund (1911) and accepted by Hutchinson (1957). In the middle thermocline Zone temperature drops at least 1°C with each 1 m increase in depth (Figure 7.4). Thus, with a typical direct stratification there are three lakes in one, an upper lake, the epithelimnion, a middle lake, the metalimnion or mesolimnion and a lower lake, the hypolimnion.

According to Hutchinson (1957) the thermocline, is not a broad region but an imaginary plane within the lake. It is located at a level intermediate between the two depths where the temperature decrease is greatest. In Figure 7.5 the strong drop in temperature of the thermocline region in the hot months of summer has been shown on the right side. If the thermocline is below the range of effective light penetration (*i.e.*

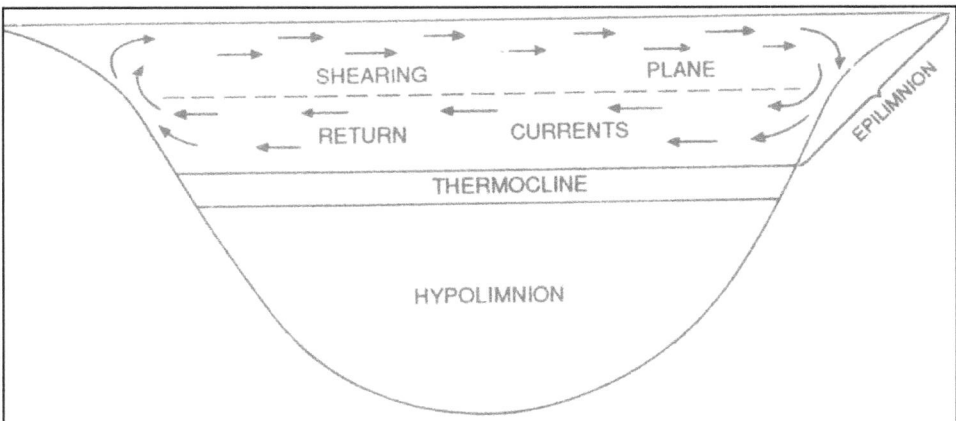

**Figure 7.3: Character of Water Circulation in the Epilimnion and Formation of a Shearing Olane.**

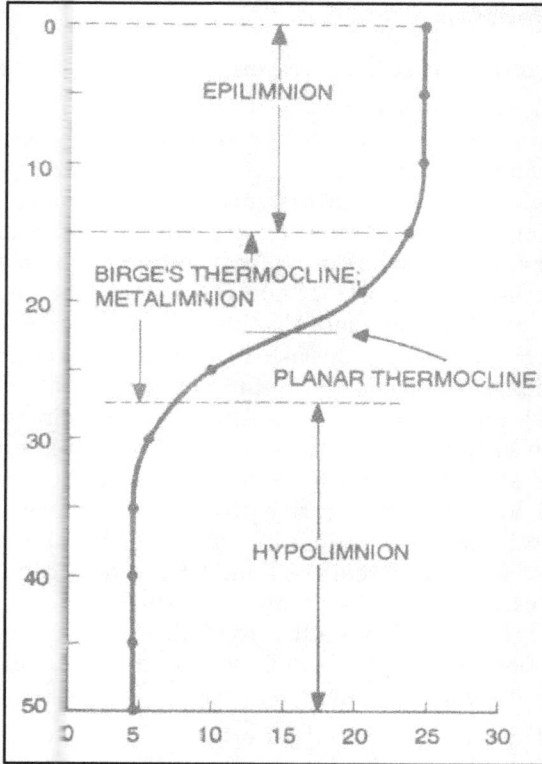

**Figure 7 4: Vertical Temperature Profile Showing Direct Stratification and Lake Region Defined by it.**

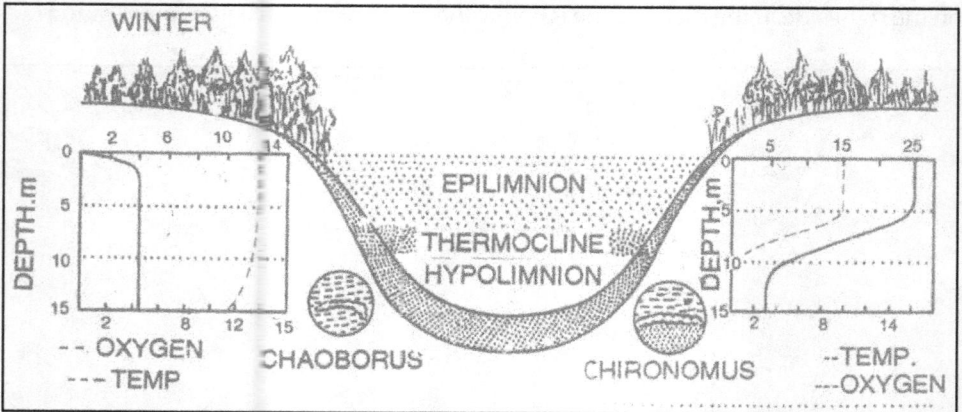

**Figure 7.5: Thermal Stratification in a North Temperate Lake.**

Summer conditions are shown on the right winter conditions on the left. In summer a warm oxygen rich circulating layer of water, the epilimnionis separated from the cold oxtgen poor hypolimnion water by a broad zone, called thermocline, which is characterized by a rap d change in temperature and oxygen with increasing depth). (Modified after Odum, 1971).

compensation level) as is often the case, the oxygen supply becomes depleted in the hypolimnion, because both the green plant and the surface source is cut-off.

With the onset of cooler weather, the temperature of epilimnion drops until it is the same as that of the hypolimnion. The water of the entire lake begins circulating during the "full overturn". As the surface water cools below 4°C, it expands, becomes lighter, remains on the surface and freezes and bring winter stratification. In winter the oxygen is usually not greatly reduced because bacterial decomposition and respiration of organisms are not so much at low temperatures and water holds more oxygen. An exception to this generalization may occur when snow covers the ice and prevents photosynthesis, resulting in oxygen depletion for the entire lake.

This leads o "winter kill" of the fish.

In the spring, as ice melts and water becomes warmer, it becomes heavier and sinks to the bottom. Thus, when the surface temperature rises to 4°C, the lake so to say takes another "deep breath" with the spring overturn. The classic pattern of two seasonal overturns is typical for many lakes in America and Eurasia. In general, the deeper the lake, the slower is the stratification and thicker the hypolimnion.

The extent of depletion of oxygen in the hypolimnion during summer stratification depends on the amount of decaying matter and on the depth of the thermocline. Productivity rich europhic lakes are subjected to greater oxygen depletion during the summer than poor obligotrophic lakes, because the 'rain' of organic manner from the limnetic and littoral zones into the profundal zone is greater in the europhic lakes. It has been found that cultural eutrophication hastens oxygen depletion in the profundal zone. Thus fishes which are stenothermal, and low temperature tolerant can survive only in less productive lakes in which cold bottom waters do not become depleted of oxygen. Lower organisms in contrast to fish of the profundal zone are adapted to withstand oxygen deficiency for appreciable periods.

If waters of a lake are very transparent and permit growth of phytoplankton below the thermocline (in the upper part of the hypolimnion), oxygen may be present here even in greater abundance than on the surface because cold water holds more oxygen. Thus the euphotic or trophogenic zone does not necessarily coincide with the epilimnion.

## Thermal Stratification in Tropical Lakes

Surface temperature of subtropical lakes never fall below 4°C. They show a distinct thermal gradient from top to bottom but experience only one general circulation period in winter. Tropical lakes as studied by Ruttner (1931) in Java, and Sumatra and reservoir by Datta Munshi and Datta Munshi (1985) have shown high surface temperature exhibiting thermal stratification in the columns of water in summer. Though the temperature gradient from surface to bottom is little in comparison to temperate lakes but this is enough to make density difference of water resulting to thermal stratification. Thus instead of the term "thermocline" it will be more appropriate to call the middle part of lake as metalinion (Verma and Datta Munshi, 1987).

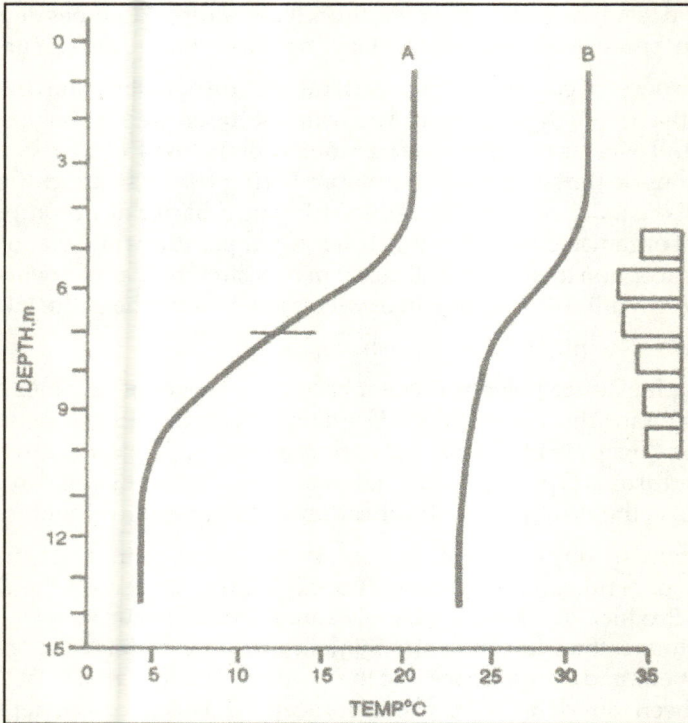

**Figure 7.6: Comparison of the Temperature Profile of a Cold Lake A and a Warmer Lake B having Identical Density Gradients Caused by Temperatures. Horizontal lines in curves represent planner thermocline based on maximum temperature change. It has also been found that very deep tropical lakes tend to remain only partly mixed *i.e.*, they are thermally stable.**

## Lake Classification of Lakes in Terms of Water Circulation

i) Dimictic (mictic-mixed). Two seasonal periods of free circulation, or over turns.

ii) Cold *monomictic*. Water never above 4°C (polar regions); seasonal overturn in summer.

iii) Warm *monomictic*. Water never below 4°C (Warm temperature subtropical); one period of circulation in winter.

iv) *Polymictic.* More or less continually circulating with only short, if any stagnation periods (high altitude, equatorial).

v) *Oligomictic.* Rarely (or very slowly) mixed (thermally stable) as in many tropical lakes.

vi) *Meromictic.* Chemically stratified lakes. Meromictic denotes partly mixed. In contrast to most lakes in which bottom and surface waters mix-periodically (*i.e.* holomictic or wholly mixed lakes) some lakes become permannnently stratified by the intrusion of saline water or salts liberated from sediments, creating a permanent density difference between surface and bottom waters. In this the boundary between circulating and non-circulating layers is a "chemocline", instead of a thermocline. Free oxygen and aerobic organisms will be absent in bottom waters of such lakes. Big soda lake in Nevada and the Hemmelsdorfersee in Germany are examples.

## Oligotrophic-eutrophic Series of Lakes Based on Productivity

Thienemann (1925) observed that conditions in deep lakes differed from those of the shallow lakes, even though they lay in the same climatic and edaphic region. The summer oxygen supply in the hypolimnion of a deep basin of oligotrophic lakes in contrast to eutrophic ones remain always abundant to support bottom fauna of diverse nature. By contrast the dissolved oxygen becomes critical in the shallower eutrophic lakes and only a few species are found in the profundal benthos (Figure 7.7). Features contrasting oligotrophic and eutrophic lakes are given below:

| *Oligotrophy* | *Eutrophy* |
|---|---|
| 1. Deep and steeped-banked | 1. Shallow, broad littoral zone |
| 2. Epilimnion volume relatively small as compared to hypolimnion | 2. Epilimnion/hypolimnion ratio greater |
| 3. Blue or green water, marked transparency. | 3. Green to yellow or brownish green, limited transparency. |
| 4. Water poor in nutrients | 4. Nutrients and $Ca^{++}$ abundant. |
| 5. Sediments low in organic matter | 5. Sediments high with organic matter. |
| 6. Oxygen abundant at all levels at all times | 6. Oxygen depleted in summer hypolimnion. |
| 7. Littoral plants limited, rosette type. | 7. Littoral plants abundant |
| 8. Phytoplankton quantitatively poor | 8. Abundant phytoplankton |
| 9. Water blooms of blue-greens lacking | 9. Water blooms are common. |
| 10. Profundal bottom fauna diverse | 10. Profundal benthos species poor |
| 11. The ratio E/H (Epilimnion / Hypolimnion) small. | 11. The ratio F/H (Eplimnion / Hypolimnion) large. |

Rodhe (1969) suggested different rates of inorganic carbon fixation by photosynthesis in these lakes. According to him 7-25gC/m$^2$/yr of lake surface indicate oligotrophy; 75 to 250gC/m$^2$ year indicate natural eutrophic lakes and 350 to 700gC/m$^2$ year in polluted lakes. Mesotrophy lies somewhere between annual primary-production of 25 to 75g carbon fixation. But extremely low rates of carbon fixation are also known in certain oilgotrophic lake. Goldman and other (1967) reported about 0.014g carbon per square metter for an average summer day in an Antartic lake.

## Special Lake Types

Seven special types of lakes based on origin have been recognized.

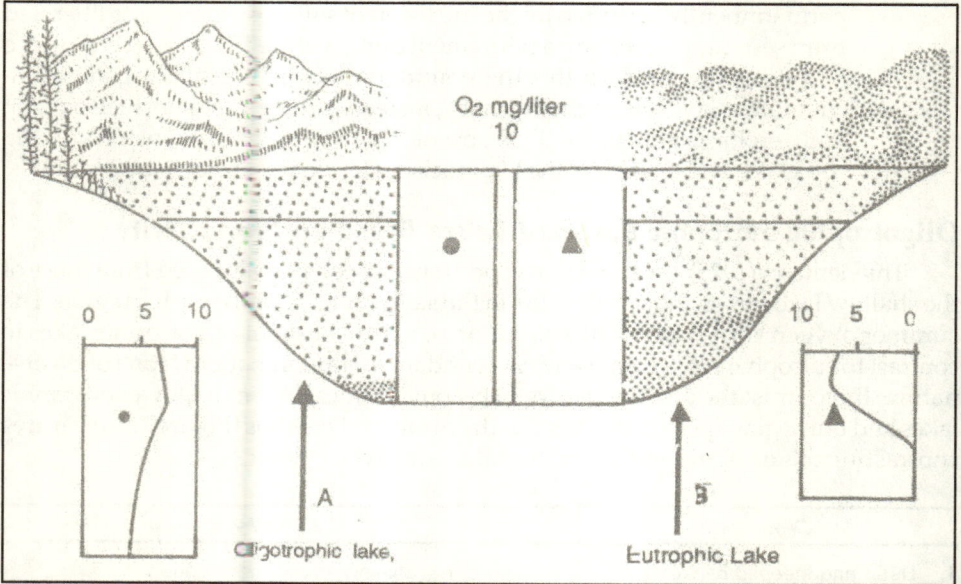

**Figure 7.7: Two Morphologically different Types of Lakes, Oligotrophic Lake at Left, Europhic Lake at Right. Epilimnion volumes and sestonic particles (dark circles) same in both lakes; hypolimnion volumes differ and oxygen profiles differ. Oligotrophic E/H volumes ratio small, eutrophic E/H volumes ratio larger. (Modified after Cole, 1979).**

## Dystrophic Lakes

These lakes generally have high concentrations of humic acid in water; bog lakes which are dystrophic in nature with low pH develop slowly into peat bogs and marshy wetlands.

## Deep Ancient lakes with Endemic Fauna

Lake Baikal in Russia is the most famous of ancient lakes. It is the deepest lake in the world and was formed by earth movement during Mesozoic era. Ninety-eight per cent of 384 species of arthropods are endemic (found nowhere else), including 291 species of amphipods, 36 species of fish are endemic. This lake is often called the "Australia of freshwater", because of its endemic fauna (Brooks, 1950).

## Desert Salt Lakes

These lakes occur in arid climates where evaporation exceeds precipitation resulting in salt concentration. The salt Lake, Utah harbour community composed of a few species which are able to tolerate high salinity. Brine shrimps (*Artemia*) are characteristic animals of such lakes.

## Desert Alkali Lakes

They occur in igneous drainages in arid climates with high pH and concentration of carbonates. Example: Pyramid Lake, Nevada.

## Volcanic Lakes

They are acid or alkaline lakes associated with regions of active volcanic activity. They have restricted biota. Examples: Some Japanese and Phillippine lakes.

## Metromictic

Chemically stratified lakes, not very common.

## Polar Lakes

Surface temperatures remain below 4°C, or rise above it for only a brief period during the summer when circulation of water can take place. Plankton population grow rapidly during this period.

# Vertical Distribution of Oxygen in Lakes

There are two main sources of oxygen in a lake, (i) by the process of photosynthesis and (ii) by diffusion of atmospheric gases at surface water. A deep stratum is away from both of these sources, and light does not penetrate into it. As a result, respiration and decomposition processes remove oxygen from the water.

In dimictic lakes during the spring and autumnal overturn of water oxygen is distributed more or less uniformly from top to bottom. If a curve is plotted based on oxygen values in relation to depth, the line will be nearly straight. This is an orthograde curve.

When thermal stratification occurs during summer months and the lake is no longer homogenous throughout, the tropholytic zone (hypolimnion) becomes isolated from the upper waters. Now oxygen begins to be consumed there. In lakes with large hypolimnion volumes and relatively little production of organic matter in the epilimnion above, the demands on the oxygen in the tropholytic zone are so slight that it shows no appreciable decline. The summer time oxygen profile, therefore, is orthograde despite thermal stratification. This is characteristic of oligotrophy (Figure 8.8). The biomass, the ratio of epilimnion volume to hypolimnion volume, and the hypolimnion temperature interact to produce vertical oxygen curves.

If environmental factors favour the production of a large epilimnion biomass of flora and fauna, the situation will be quite different during summer stratification. Great quantities of dead and dying organic matter consume large amount of oxygen in hypolimnion water. Thus oxygen curve shows sharp decline from top to bottom. So the vertical curve (orthograde) in periods of circulation changes to elinograde curve (Figure 7.8). Clinograde oxygen distribution characterizes stratified eutrophic lakes.

# Important Reservoirs of India

The most important reservoirs of India are constructed mainly by impounding river systems which are as follows:

   a) Reservoirs of the Ganga river system. These are Rihand, Matatila, Sardasagar in Uttar Pradesh; Tilaiya, Maithon, Mayurakshi, Panchet in Bihar; Kangsabati in West Bengal; and Govind sagar in Madhya Pradesh.

b) Reservoirs of the Indus river system – In Punjab and Himachal Pradesh Beas Dam and Govindsagar are considered to be very important reservoirs.

c) Reservoir of Mahanadi river system – In Orissa (Sambalpur), one of the longest dam of the world called Hirakund is situated. Reservoirs of Krishna river system – Nagarjunsagar dam and Nizamsagar dam in Andhra Pradesh and Tunga Bhadra in Mysore State are very important reservoirs of this river system.

d) Reservoirs of the Cauvery river system – Krishnarajasagar in Mysore and Bhavanisagar and Mettur dam (Stanley reservoir) in Tamil Nadu are the important reservoirs of this river system.

The commercially important species of fishes in these reservoir fisheries are *Labeo rohita, L. bata, L. calbasu, L. fimbriatus, Catla catla, Cirrhinus mrigala, Mystus seenghala, Bagarius bagarius*, etc.

# Case Study of a Reservoir of India

## Location and Hydrography

Badua reservoir is located 80 km south from Bhagalpur town (86°30 E longitude and 25°30'N latitude) in Bihar state of India. It has been constructed by putting an earthen dam across a seasonal stream 'Badua'. It receives rain water from its vast catchment areas of 296.96 sq. km. It has been commissioned in 1963 for irrigation and fishery purposes. Maximum air temperatures (33.6-37.1°C) were recorded during summer (April-June) and minimum (21.7-24.0°C) during winter months (December-January). Heavy rainfall was recorded during July-October (163.5-468.0mm) (Verma and Datta Munshi, 1983; Datta Munshi and Datta Munshi, 1988).

The hydrographic particulars are given below:

| | | |
|---|---:|---|
| Reservoir area | 8.80 | km² |
| Catchment area | 296.96 | km² |
| High flood level | 133.59 | m² |
| River bed level | 94.55 | m² |
| Maximum depth | 32.00 | m² |
| Mean depth | 14.57 | m² |
| Gross capacity | 128.29 x 10⁶ | m² |
| Live capacity | 100.77 x 10⁶ | m² |
| Dead storage | 18.50 x 10⁶ | m² |
| Discharge of right canal | 550 | m²/S |
| Discharge of left canal | 450 | m²/S |
| Net irrigable area | 424.9 | Km² |

## Thermal Stratification

A vertical thermal gradient was observed during summer and resulted in the

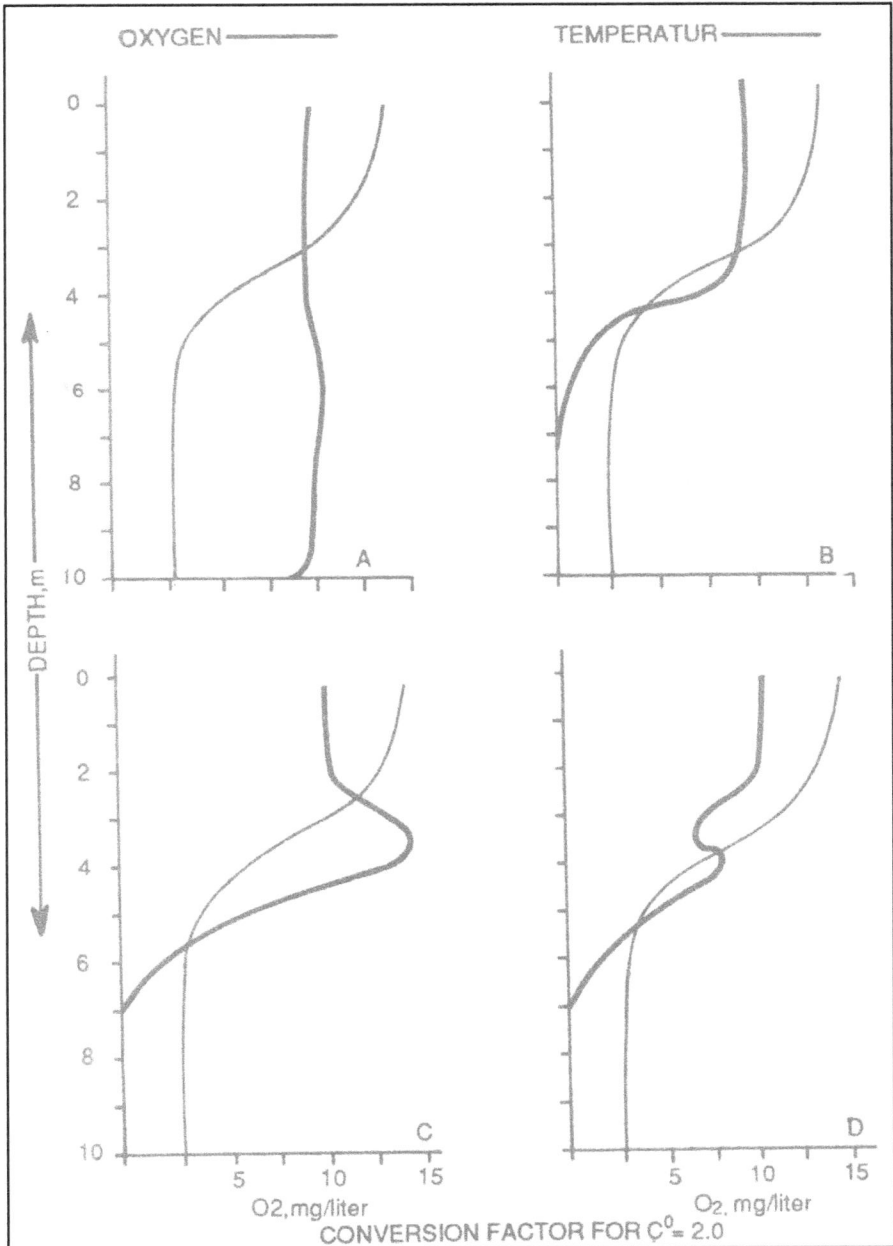

**Figure 7.8: Oxygen Profiles in Thermally Stratified Lakes.**

A: Orthograde oxygen profile; B. Clinograde oxygen profile; C: Positive heterograde oxygen profile; D: Negative heterograde profile. Oxygen plotted with thicker lines; temperature plotted uniformly and with thinner lines. Figures on X axis times 2.0-0°C (Modified after Cole, 1979).

formation of three distinctive thermal strata. From surface to 5m depth, the temperature was uniform (30.0-29.0°C) which represents the Epilimnion. Again at the bottom below 10m depth there was uniformity in the temperature (25.9-24.0°C) representing the Hypolimnion. In between these two regions there was a sharp fall in temperature (29.0 to 25.0°C) which represents the Metalimnion (Figure 7.9). (Verma and Datta Munshi, 1987).

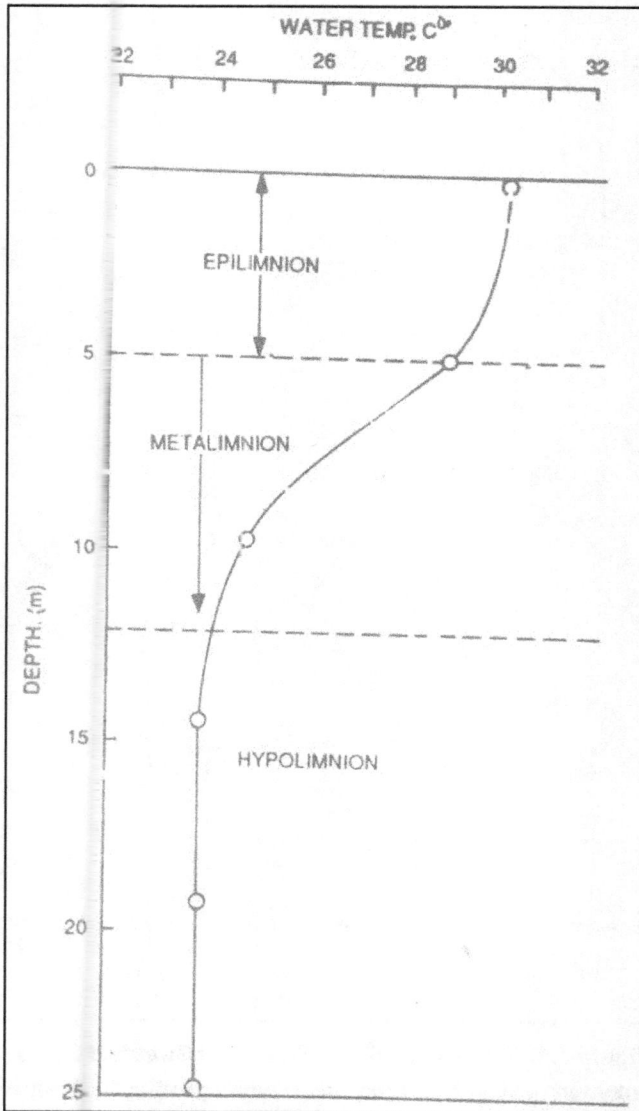

**Figure 7.9: Water Temperature Profile Indicating Thermal Strata during Summer Season.**

Heavy rainfall (169-458 mm) and strong wind velocity (6.2-10.4 km/h) during the monsoon months, lowered the temperature of the epilimnion and the thermal stratification thus disappeared. The temperature of the water column develops only 1°C difference from surface to bottom. It leads to an isothermal condition in monsoon and winter months (Figure 7.9).

## pH Valve and Oxygen Content

The medium was always in the alkaline range (7.5-8.1), pH value graduality decreased from surface to bottom and the fall was remarkable in summer sampling where a difference of 0.35 was recorded (surface pH=7.90; bottom pH=7.55). During the monsoon, the value changed at different depths from 7.7 to 7.5 but in winter fluctuation from surface (7.1) to bottom (8.0) was small.

Dissolved oxygen was distributed more or less uniformly from top to bottom showing an orthograde profile (Reid and Wood, 1976) throughout all the seasons except in summer. The entire water column was more than 50 per cent saturation of oxygen and provide a suitable habitat of fish. In Winter high saturation values (surface- 94 per cent bottom- 90 per cent) were obtained.

## Free Carbon Dioxide

Free carbon dioxide increased slightly with depth having maximum value 1.5-2.0 mg/l) in the hypolimnion zone. During summer (April-May) the epilimnion (0.2m) was typically devoid of free carbon dioxide, but contained carbonate.

## Alkalinity

Carbonate alkalinity was only recorded during summer in the epilimnion zone (4.0-4.5 mg/l). Bicarbonate alkalinity was recorded in all the seasons. Its value at different depths ranged between 75.5-71.5 mg/l during the summer, 74.0-73.9 mg/l in the monsoon and 68.5-68.0 mg/l in the winter, thus showing a minor gradient with depth.

## Nutrient Status

Silica value showed a regular trend of variation in all the seasons ranging from the lowest value 23.0 mg/l in the winter to 33.0 mg/l during the summer. Phosphate value showed a gradient with depth reaching its peak below a depth of 15m. From surface to bottom the value ranged between 0.214-0.321 mg/l in the summer, 0.093-0.160 mg/l in the monsoon and 0.067-0.187 mg/l in the winter season. Nitrate content was also maximum near the bottom (10-25m) ranging from 0.40 mg/l at the surface to 0.85 mg/l at the bottom in the summer, 0.56 mg/l at the surface and 0.73 mg/l at the bottom in the monsoon and 0.24 mg/l at the surface and 0.66 mg/l at the bottom in the winter season. Chloride content was very low and did not exhibit depth gradient. (Figure 7.10).

## Primary Productivity

Productivity value was maximum at 0.5 meter where gross and net values were 0.0980-0.0734 gC/m$^2$/hr and 0.074-0.0572 gC/m$^2$/hr respectively. The lower limit of productivity was at 2.0 m depth where gross productivity values were 0.0188-0.0107

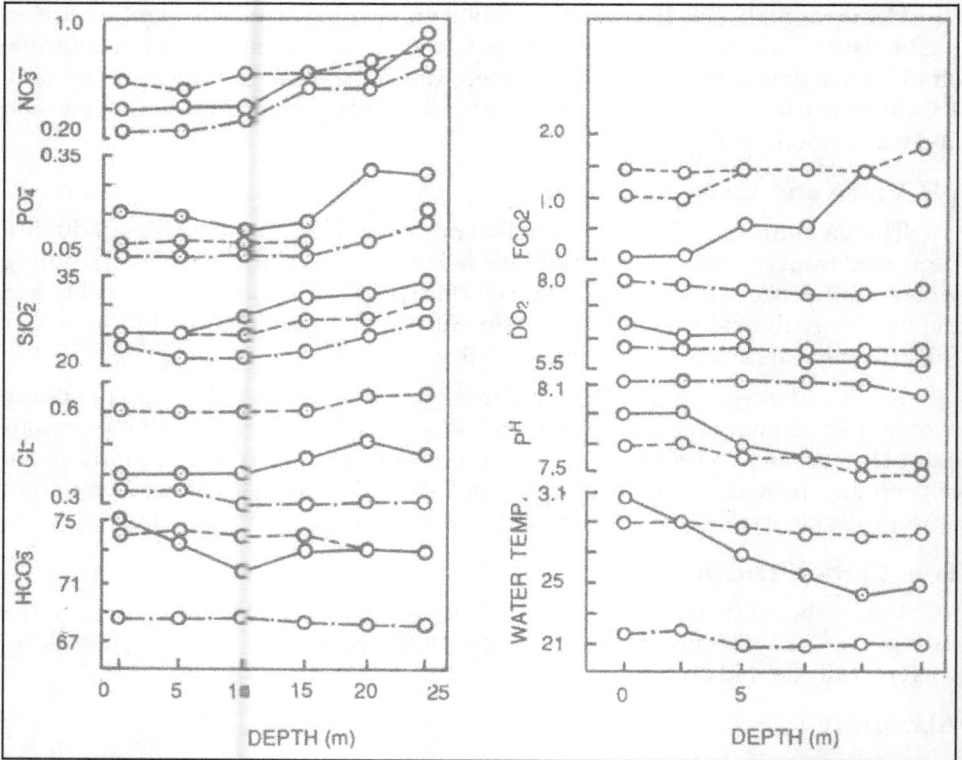

**Figure 7.10: Seasonal Variation in Vertical Distribution of Water Temperature, pH, Dissolved Oxygen (DO₂), Free Carbondioxide (FCO₂), Bicarbonate Alkalinity (HCO₃), Chloride (Cl⁻, Silica (SIO₄), Phosphate (PO₄) and Nitrate (NO₃) in mg/l – Summer, — Monsoon and – Winter.**

gC/m³/hr, net productivity as determined per square meter water surface area was comparatively higher during April than in November (Figure 7.11). The light intensity was at its peak during summer months.

## Plankton

The phytoplankton community constituted 64.9 to 66.8 per cent of the total plankton and showed its peak in January and September. The members belonging to Bacillariophyceae (80.5 per cent) and Cyanophyceae were mainly represented by *Cymbella turgida, Synedra ulna, Gomphonema sphaerophorum, G. lanceolatum* and *Navicula cuspidate. Microcystis aeruginosa* alone formed the dominant member of Cyanophyceae. Other forms recorded were *Aphanocapsa elechista, Merismopedia glauca, Anabaena unispora* and *Gloeocapsa punctata*. Members of chlorophyceae *viz., Chlorococcum* sp., *Closterium setaceum, Cosmarium puntulatum, Bulbochaere oblicua* and *Oedogonium oriforme* appeared during the summer months (April-May).

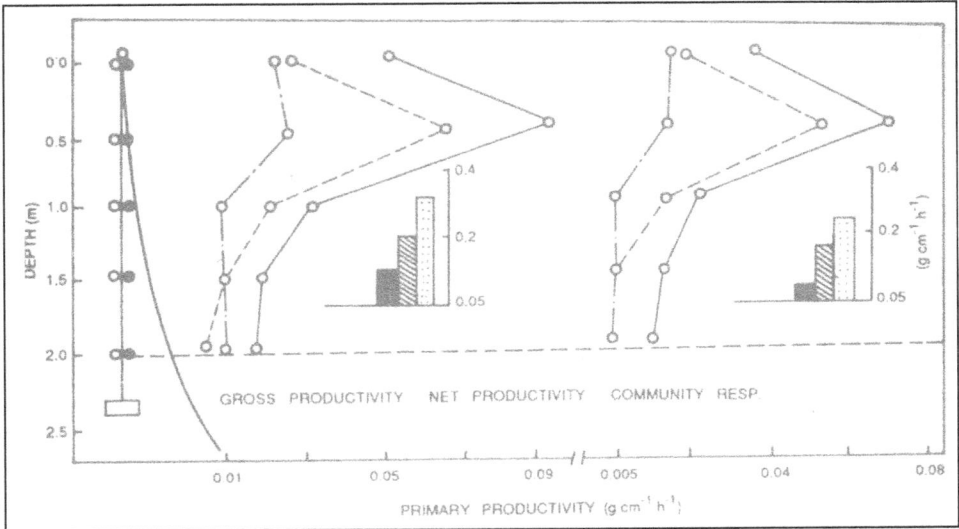

**Figure 7.11: Measurement of Primary Production with Light and Dark Bottle a different Depths. Vertical bar in lower right corner indicates total production (gC/m²/h) in two meter water column.**

Zooplankton community constituted 33.2 to 35.0 per cent of the total plankton hauled. Its bimodal peak in February and December was dominated by rotifers (25.1 per cent) and Copepods (84.2 per cent) respectively (Figure 7.12). The rotifers were mainly represented by *Keratella tropica, Filinua terminalis, Bracnionus calyciflorus, Lecane luna* and *Polyasthra* sp. Copepods formed the dominant groups among zcoplankton community and were represented by *Mesocyclops lenckart, Mesocyclops hyalimus, Heliodiaptoms viguus, Rhineodiapfomus indicus* and *Spicodiaptomus* sp. Few Cladocerans represented by *Moina mucruha, Diaphanosoma excisum, Ceriodaphnia* sp., *Chydorus* sp. and *Alonella dentifera* were recorded in July-August. Rhizopods recorded were represented by *Centrojysix ecornis, Arcella vulgaris, Difflugia corona* and *Englypha tubericulate*. The phyto-and zooplankton showed their maximum concentration up to 2.0 m depth but became limited at 10-15m.

## Fish Production

The fisheries department have regularly stocked the reservoir with fingerlings of *Catla catla, Lebeo rohita* and *Cirrhinus mrigala* in the ratio of 40:30:30 from 1975 onwards. An estimated number of 11,28,500 fingerlings of these carps were released in the reservoir till 1979. From the average length frequency and number of representative species of the fingerlings released it was possible to determine instantaneous mortality (i) = 0.7911, survival rate (s) = 0.5913 and annual mortality (r) = 0.4087. The population structure to stocking are furnished in (Table 7.1).

Fishes captured were healthy. The average weight of these ranged from 3.5-12.1 kg for Catla, 1.5-6.6 kg for Labeo and 1.6-9.7 kg for Mrigal. The annual fish production ranged from 0.04-0.07 g/m² (Tables 7.2 and 7.3).

**Table 7.1: Estimation of Fish Population and Production in Relation to Stocking in the Reservoir**

| Year | Fingerling Released (Nos.) | Fingerling Survived (Nos.) | Annual Catch (Nos.) | Fishes Remains in the Basin 78-79 | Fish Population 78-79 | | Fish Landing 78-79 | | |
|---|---|---|---|---|---|---|---|---|---|
| | | | | | No. in Total | Weight# | Numbers | Weight (kg) | Cost (Rs.) |
| **Catla** | | | | | | | | | |
| 1975-76 | 1,54,600 | 63,849 | 1594 | 62,255 | | | | | |
| 1976-77 | 1,32,000 | 54,516 | 2394 | 52,182 | 1,81,0032 | 1,37,584 | 2344 | 2729.6 | 19,107 |
| 1977-78 | 1,36,000 | 56,168 | 1417 | 54,751 | | | | | |
| 1978-79 | 28,800 | 11,804 | – | 11,894 | | | | | |
| **Rohu** | | | | | | | | | |
| 1975-76 | 1,16,950 | 47,807 | 1355 | 46,532 | | | | | |
| 1976-77 | 90,000 | 40,887 | 1811 | 39,576 | 1,35,886 | 5,02,778 | 1392 | 1268.3 | 8.878 |
| 1977-78 | 1,02,000 | 42,126 | 1268 | 40,858 | | | | | |
| 1978-79 | 21,600 | 8,920 | – | 8,920 | | | | | |
| **Mrigal** | | | | | | | | | |
| 1975-76 | 1,15,950 | 47,887 | 849 | 47,038 | | | | | |
| 1976-77 | 99,000 | 40,887 | 1210 | 39,677 | 1,36,860 | 8,07,474 | 1333 | 1877.1 | 13,140 |
| 1977-78 | 1,02,000 | 42,126 | 901 | 41,225 | | | | | |
| 1978-79 | 21,600 | 8,920 | – | 8,920 | | | | | |

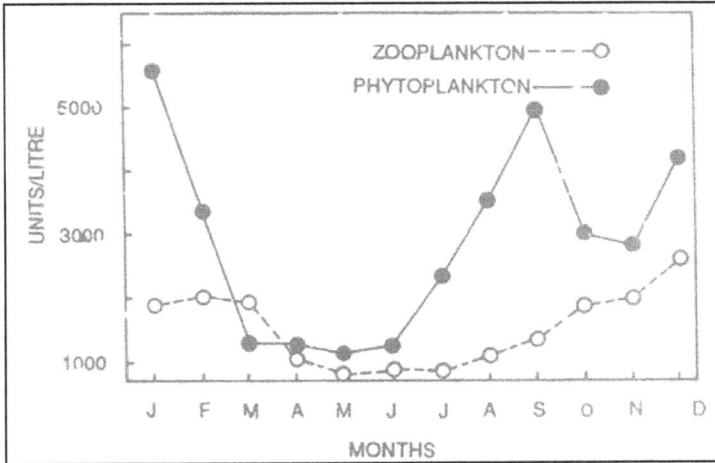

**Figure 7.12: Monthly Variation in Total Number of Phytoplankton and Zooplankton.**

**Table 7.2: Length and Corresponding Weights of Fishes Captured**

| Fishes | Length (mm) | Weight (g) |
|---|---|---|
| *Catla catla* (Ham) | 610–864 | 3550–12100 |
| *Labeo rohita* (Ham) | 432–762 | 1500–5600 |
| *Cirrhinus mrigala* (Ham) | 538–889 | 1630–9700 |

**Table 7.3: Annual Fish Landings and Rate of Fish Production**

| Year | Fish Landings (Quintal) | Fish Productiong/m² |
|---|---|---|
| 1975-76 | 40.3 | 0.0498 |
| 1976-77 | 39.0 | 0.0482 |
| 1977-78 | 47.0 | 0.0583 |
| 1978-79 | 58.8 | 0.0726 |

## Modelling and Management (Energy Budget)

Models particularly play an important role in synthesis by providing mechanisms for combining diverse experimental results and integrating them with the literature. Model which incorporate various components of the system that can be quantified in biomass, nutrients, energy etc. help in simplifying and giving an unified picture of the system. But this can never be exact replicas of the system itself, they are infact abstractions.

Graphic models of the reservoir ecosystem have been prepared to give a unified energy flow picture through different trophic levels in each of the system (Figure 7.13). The quantum of solar energy entering the earth's atmosphere is approximately

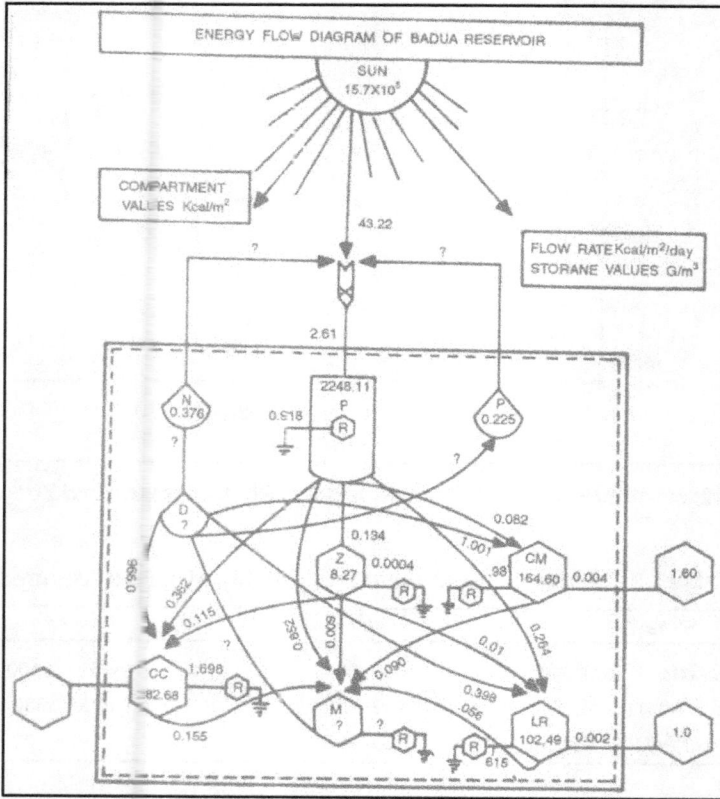

**Figure 7.13: Conceptual and Diagrammatic Model of Energy Flow through different Trophic Levels of the Reservoir Ecosystem.**

$15.7 \times 10^5$ cal/m²/yr. Much of the energy is reflected back due to the presence of cloud, dust area per unit time actually available for photosynthesis varies with geographical locations which comes in the order of $1.5 \times 10^4$ kcal/m²/yr in these systems.

From the graphic model it is clear that the rate of the primary production was 2.613 kcal/m²/day and energy loss due to respiration was 0.918 kcal/m². the nutrient nitrates (0.376 g/m²) were determined.

The secondary production of zooplankton which graze upon phytoplankton was at the rate of 0.1349 kcal/m²/day). The fishes also ingest phytoplankton directly and the energy flowing through this link has been calculated (Catla catla 0.362 kcal/m²/day). Energy loss through respiration of these fishes were 619.86, 224.73 and 360.93 kcal/m² for Catla, Labeo and Mrigala respectively. However some energy was dispersed in the system through faecal matter and waste products of fishes, which when calculated specieswise comes to 55.53 kcal/m² for Catla, 20.49 kcal/m² from Labeo and 32.92 kcal/m² for Mrigala. Net energy available for human consumption through annual catches was 3.66 kcal/m² from Catla,1.05 kcal/m² from Labeo and

1.60 kcal/m$^2$ from Mrigala. The locked up energy that remained in the form of fish biomass in the basin has been calculated species wise, *Catla catla* 282.68 kcal/m$^2$, *Labeo rohita* 102.49 kcal/m, and *Cirrhinus mrigala* 164.60 kcal/m$^2$ (Figure 7.13).

# 8

# Ponds

For the definition of ponds as opposed to lakes the depth of water is of first importance than the actual size of the water surface. The depth of a pond is rarely more than 2m. The biological processes in a pond are more intensive and rapid than a deeper lake. The littoral zone of ponds is relatively larger compared to that of lakes. In ponds limnetic and profundal zones are small or absent and hence stratification has little importance. In contrast to the limnetic zone of a lake the littoral zone is the chief producing region in ponds. The communities in this zone are of prime interest. Ponds are present in those areas where there is adequate rainfall and they are mostly rainfed.

## Classification of Ponds

### Natural Ponds

Natural ponds are perennial shallow water bodies. These are formed in various ways. When a stream shifts its position it leaves behind an isolated body of standing water which forms the "Ox-Bow" pond. In limestone regions where depressions are formed due to the solution of the underlying strata, the water gets accumulated either by flood water or rainfall and natural ponds are formed. Sometimes the last remnant of a lake whose basin has become filled progressively by sedimentation in course of time is transformed into a pond.

### Fish-Pond

Most of the fish-ponds are semiartificial ponds. Some are constructed by erecting dams across a stream or basin by man. With few exception their water level can be regulated by inflow and drainage where pisciculture is practiced. Fish pond is a shallow body of water that can be drained completely. It is often supplied by running water, but also by spring, ground or rain water. When a pond is formed, it contains water periodically. Naturally, the change from wet to dry conditions as well as various cultivation processes closely affect the flora and fauna of a fish pond. Generally little water is allowed to flow through them.

## Beaver Ponds

These ponds are created by making holes in bank of streams passing through forests by beavers. Beaver ponds were common in North America in primevaldays. A beaver pond is abandoned when the supply of plants in the vicinity become reduced. Beaver ponds act as water reservoirs.

## Special Types of Ponds

Apart from the above mentioned ponds there are duck breeding ponds, and the various types of construction by the Pollution Control Board authorities as sewage, oxidation, stabilization, maturation and settling ponds. Some ponds are kept for ornamental purposes.

### Pools or temporary Ponds

They occur in depression in the ground either at the margin of glaciers where they fill with meltwater or in the vicinity of river bed, after the floods have receded. The water thus collected usually is very shallow and measures maximum to a few feet only. Also prolonged rainfall may form temporary small pools. All these pools dry up in some part of the year, and as such organisms in these habitats must be able to survive in a dormant sage during dry periods and be able to move in and out of the pools like amphibious and adult aquatic insects. Various types of flora and fauna of a pond have been described.

## Case Study of a Fish – Pond

A typical fish-pond has been chosen for the present study (Datta Munshi, 1974, Chakravarty and Datta Munshi, 1975, Premkumar and Datta Munshi, 1977).The pond lies in the city of Bhagalpur (latitude 25°15′ north and longitude 87°02′ east)Bihar, India. A contour map of the pond has been constructed in summer season with the help of bamboo poles. The depth of water having similar heights are connected with the help of ropes for preparation of the contour map of the pond. The pond is almost square in shape having an area of approximately 0.84 hectare. The depth of water in the summers season ranged from minimum 0.2 near the margin to maximum 0.91 m in the centre.

### Physico-chemical Properties of Pond Water

The mean water temperature in the surface layer varied from 19.5°C to 35.0°C during the period of the study attaining the maximum in September and the minimum is January. The transparency was maximum in April (92.0 cm) showing highest light penetration in the aquatic body. pH of the pond was always in alkaline side. High dissolved oxygen values were observed during July (10.0 ppm), August (16.7 ppm) and September (15.0 ppm). During this period the pond got infested with aquatic macrophytes and in the month of August maximum biomass of these plants (451.2 g dry wt/m$^2$) were obtained (Nasar and Datta Munshi, 1972). Free carbon dioxide was mostly nil in this pond except in winter months. It may be assumed from the free $CO_2$ and carbonate alkalinity data that except in the months of October,November and December, there is always conversion of the dissolved carbon dioxide to calcium carbonage in the pond. The bicarbonate alkalinity of the pond water was 343.0-387.0 ppm) quite high during March to June showing high metabolism of the pond.

### Algal Blooms

Blooms of Algae occurred in the summer season which continued upto rainy season. The blooms of *Spirogyra* were observed in the months of March andApril followed by blooms of *Oedogonium*. Diatoms (Bacillariophyceae) predominated in February-March and again in June-July. The blooms of phytoflagellates comprising species *Trachelomonous* and *Lipocynclis* were observed in the rainy season. It was interesting to note that *Anabaena* bloom remained in the pond from last weak of October to 1$^{st}$ weak of December forming a thick carpet like structure.

### Zoo-biota

### *Microfauna (Copepods. Cladocerans, Rotifers)*

The zooplankton exhibited seasonal periodicity in their abundance similar to that of phytoplankton. In ponds with thick deposits of organic matter, there was an abundance of cladocerans, but fewer copepods and rotifers. The zooplankton organisms are of general importance in the productivity of fish ponds, as they form the main food of young fish.

### Macrofauna

Oliogochaetes, such as *Nais, Chaetogaster, Dero, Autophorus* were found among vegetation in the bottom mud.

Crustaceans were represented by prawns, shrimps and crabs.

### Insect

Both adults and larval forms of Mayflies (*Ephemeroptera*), *Caddisflies*(Trichoptera), Midges (*Diptera*), Mosquito larvae, *Chirononids,* water bugs like *Notonecta, Nepa, ranatra;* and *Cybister-Beetles* were abundant.

Other macrofaura were represented by fishes, frogs, snakes and birds.

## Primary Production by Algae

The data for monthly variation of the primary production by algal component was determined with the help of day light and dark.

It was observed that the net primary production of algae (Phytoplankton and filamentous forms) was maximum in the months of April (3.50 gC/m$^2$/day). The study revealed that the maximum value was seven times that of minimum value of net primary production. In tropical waters preduction is moderate throughout the year with little variation. This is in contrast to the temperate waters where production of springs and summer may be fifty times or more than in autumn and winter (Rakmount, 1966).

The high net primary productivity in March and April 3.03 to 3.50 gC/m²/day might be due to higher transparency leading to more penetration of incident light to water and optimum temperature for the photosynthesis of algal producers.

## Macrophyte Biomass Production

In this fish pond the aquatic macrophytes were weeded out to some extent in the early summer to clear the habitat. Spawn and fry of desirable carp fish were introduced into the pond water in the month of August-September. In rainy season the existing macrophyte vegetation had grown profusely attending maximum biomass of 4512 g/m² in the month of August.

## On the Algal Flora of some Ponds of Bhagalpur, India

The value of plankton and other algae as direct or indirect food for fish, their usefulness as indicators of water conditions and the harmful effects caused by their excessive development in fishery waters and waters supplies have been recognized not only by fishery biologists and water work engineers but also by fish farmers and general public. In this context it is necessary to review the status of algal flora of inland waters. The occurrence of a large number of algae in other parts of India as reported by Philipose (1959), Singh (1961), Ahmad (1967); Nair (1967) and Gupta (1972) prompted the authors to explore the part of India.

## Seasonal Variations in the Physico-chemical and Biological Properties of a Tropical Shallow Pond

Nassar, S.A.K. and Jayashree Datta Munshi (Ecol. Res. Lab. Post Graduate Department, Zoology, Bhagalpur University, Bihar, India. Seasonal variations in the physico-chemical and Biological properties of a Tropical shallow Ponds have been investigated.

The physico-chemical factors of a tropical freshwater shallow pond in Bhagalpur (India) have been studied along with seasonal variations of plankton and discussed in the present communication (1) Basing upon the abiotic environmental characteristics and the biotic nature, the shallow pond surveyed can be juged to be one of the "eutrophic" water areas. (2) The considerable fluctuation of dissolved oxygen has been correlated with the seasonal changes of macrophytic biomass of the pond (3) The absence of free $CO_2$ and abundance of macrophytes, plankton and airbreathing fishes indicate that free $CO_2$ is not a conditioning factor for their production and growth (4) Phosphate in the pond soil was not entirely aborted by the profuse growth of macrophytes even during the intensive photosynthesis (5) The fishes found in the pond are mostly air-breathing and provided with accessory respiratory organ. Some of the weed fishes have indirect food relation with the macrophytes, because they harbor all types of live food organisms of fishes. These fishes also use the macrovegetation for breeding purposes, (6) Such type of ponds infested with microphytes is suitable for the inhabitation of air-breathing fishes.

The production of phytoplankton and fishes in deep water ponds or lakes has been studied to a limited extent in India (Ganapati, 1940, Sreenivasan 1964, Hussain 1967) but practically no work on the biological properties of shallow ponds in relation to their physico-chemical and climatic conditions in Bihar (india) has been done so far. The association of plants and fishes, as well as other organisms in quite different in shallow waters from that in the deep water ponds or lakes so far investigated.

Ganapati (1979) pointed out that no attempt has been made in tropical waters to determine the exact relationship between the biota and the habitat during different seasons of a year or years.

With a view to understand the seasonal variation in the physico-chemical and biological properties of a freshwater tropical shallow pond the present work was undertaken. The present report embodies an account of some aspects of selected physico-chemical and biological nature of a shallow pond under different climatic conditions at Bhagalpur, India together with the composition are seasonal abundance of plankton, the larger aquatic vegetation and littoral and profound fauna. The study was conducted during May, 1971 to April, 1972.

## Description of the Pond

The pond under investigations known as "Hathkatora Pond", lies west of the Post-Graduate Department of Zoology, University of Bhagalpur, Bihar (India). The city of Bhagalpur is situated on the southern bank of the Ganga River, in latitude 25°15′ north and longitude 87°02′ east. The pond is almost square in shape having an area of apptoximately 0.54 hectare, with a depth ranging from 0.30 to 0.91 m. It is evident from old records that about a hundred years ago the pond was very large and deep, it was infected with different types of macrophytic vegetation and gradually became smaller and shallower in the course of time.

## Materials and Methods

Samples of water and soil for physico-chemical analysis were conducted at fortnightly intervals. Temperature was measured with a mercury thermometer graduated with to 119°C. The hydrogen-ion concentration, dissolved oxygen, free $CO_2$ of the pond water and total soluble salts, phosphate, calcium carbonate of the pond soil were determined according to the standard methods. Samples of plankton and micro-organisms were obtained by filtering the water taken from different depths and stations in the pond, through plankton net (made of bolting silk No.21 with 77mesh/sq.cm). The concentrated plankton was then preserved in four percent formalin for subsequent qualitative determination. The macrovegetation and macrofauna were collected several times, each month, with the help of a fishing net, Ekman's dredge and plankton net. Data on temperature, rainfall and humidity were obtained from the nearest meterological station (Sabour).

## Physical Properties of the Pond

The maximum depth of the pond was 0.91 m in August and the minimum (0.30m) in June. The water in the pond remained colourless throughout the period of investigation. The mean water temperature in the surface layer varied from 22.2°C to 32.5°C during the period of this study attaining the maximum in April and the minimum in December (Table 8.1).

**Table 8.1: Physico-chemical Properties of the Pond Water**

| Month | Temp. (°C) | Dissolved Oxygen (ppm) | pH | Free Carbon Dioxide |
|---|---|---|---|---|
| **1971** | | | | |
| May | 30.2 | 6.3 | 7.5 | Nil |
| June | 30.1 | 7.3 | 8.0 | Nil |
| July | 29.2 | 7.9 | 8.5 | Nil |
| August | 28.3 | 8.9 | 8.0 | Nil |
| September | 29.1 | 7.1 | 8.5 | Nil |
| October | 28.6 | 3.2 | 8.0 | Nil |
| November | 24.3 | 3.5 | 8.0 | Nil |
| December | 22.2 | 3.8 | 7.5 | Nil |
| **1972** | | | | |
| January | 22.5 | 2.0 | 7.5 | Nil |
| February | 22.7 | 1.9 | 7.0 | Nil |
| March | 28.3 | 2.4 | 7.5 | Nil |
| April | 32.5 | 3.5 | 7.0 | Nil |

High dissolved oxygen values were observed during the period of August (8.9 þþm) and the lowest value during February (1.9 þþm).

**Table 8.2: Chemical Properties of the Pond Soil**

| Month | $P_2O_2$ (ppm) | Total Soluble Salts (μ mhos) | $CaCO_3$ |
|---|---|---|---|
| 1971 | | | |
| May | 8 | 0.396 | 0.98 |
| June | 8 | 0.283 | 1.16 |
| July | 12 | 0.294 | 1.20 |
| August | 12 | 0.297 | 1.35 |
| September | 8 | 0.254 | 1.44 |
| October | 8 | 0.243 | 1.14 |
| November | 8 | 0.290 | 1.22 |
| December | 12 | 0.305 | 1.35 |
| 1972 | | | |
| January | 12 | 0.276 | 1.44 |
| February | 24 | 0.117 | 0.98 |
| March | 8 | 0.283 | 0.61 |
| April | 12 | 0.303 | 0.98 |

Hydrogen-ion-concentration (pH) ranges between 7.0 to 8.5 with the maximum in September and October and the minimum in February and April.

Free carbon dioxide was not detected in this water body throughout the period of investigation.

The total soluble salts present in the pond soil varied from 0.117 to 0.396 μ mhos in electrical conductivity, with the maximum in May and the minimum in February. The calcium carbonate varied from 0.61 present to 1.44 present during the period of study. The phosphate as estimated, gave a value of 8 to 24 þþm in $P_2O_3$ during the period of study.

The climate at the pond site was mainly dry (29 per cent to 80 per cent humidity) during the period of the study. The average monthly mean maximum temperature ranged from 23.3°C to 37°C and mean minimum temperature varied from 7.3°C to 25.0°C. The average monthly rainfall ranged from 0.0 mm to 454.2 mm (maximum in August)

### Table 8.3: Meteorological Conditions of Bhagalpur Temperature (°C)

| Month | Mean Max. | Mean Min. | Rainfall (mm) | Humidity (per cent) |
|---|---|---|---|---|
| 1971 | | | | |
| May | 33.4 | 22.7 | 170.4 | 60 |
| June | 32.3 | 24.9 | 204.9 | 73 |
| July | 31.6 | 24.9 | 278.6 | 79 |
| August | 30.7 | 25.0 | 254.2 | 80 |
| September | 32.1 | 24.9 | 113.2 | 69 |
| October | 31.3 | 21.9 | 203.4 | 65 |
| November | 27.5 | 14.7 | 22.4 | 51 |
| December | 24.6 | 9.2 | Nil | 43 |
| 1972 | | | | |
| January | 23.3 | 7.2 | 22.4 | 46 |
| February | 23.4 | 9.2 | 40.2 | 51 |
| March | 32.3 | 14.1 | Nil | 30 |
| April | 37.3 | 30.0 | 1.8 | 29 |

## Biological Properties of the Pond

### Phyto-biota

a) The phytoplankton at different seasons were as follows:

Summer season (March to June)

Chlorophyceae – *Volvox aureus; Ceratium* spp.

Euglenineae – *Eugleno tuba*

### *Monsoon Season (July to October)*

*Chlorophyceae – Volvox auereus; Closterium acerosum; Scedesmus quadricaudata; Pondorina morum; Pediastrum simplex*

*Bacillariophyceae – Cyclotella* spp*; Symdra ulna; Navicula* spp.

*Myxophyeae – Oscillatoria chlorine; Spirogyra* spp.

### Winter Season (November to February)
*Chlorophyceae – Scenedesmus quakicoudata;*

*Euglena virdis: Pediastram simplex.*

*Bacillariophyceae – Navicula* spp., *Navicula cuspidate.*

*Myxophyceae – Microcystis aeruginosa*

## Macrovegetation
*Alternenthera sessilis: Ramunculas selerats*

*Euhydra fluctuans: Spirodela polyza; Marsilea quadrifolia*

## Zoo-biota
a)  The zooplankton at different seasons were at follows:

### Summer Season
*Crustacea – Cyclops viridis; Moina dubia; Dophnia lumholtri; Simocephalus, vetulus; Diaphanosoma sarssl.*

*Rotifer – Asplanchana priodonta; Brachiomus falcatus; B. quadridentata; B. angularis.*

### Monsoon Season
*Crustacea – Mesocyclops kyalinus; Cyclops* spp*; Spicodioptomus* spp*; Moina dubia; Bosmina longirostris; Diaphanosoma excisum.*

*Rotifer – Fillinia longiseta; Brachiomus caudatus; B. forficula; B. falcatus; Conochilus spp.*

### Winter Season
*Crustacea – Cyclops viridis; Cyclops* spp*; Diaptomus caducus; Bosmina longiseta; Daphnia carlnata; Cerodaphnia reticulate.*

*Rotifer – Asplanchna priodonia; Brachiomus calyci floras; Fullnia Longiseta*

b)  Weed dwelling fauna;

   *Pila globosa; Hirudinea* spp.

c)  Shore dwelling fauna;

   *Pheretima posthuma*

d)  Fish fauna

   *Puntius sphore; Esomus donricus; Channa punctatus; Trichogaster faciatus; Amphipnous cuchia.*

## Discussion
The freshwater algae of Bhagalpur (India) is composed mainly of the Chlorophyceae, the Bacillariophyceae, the Euglenophyceae, the Cyanophyceae and, rarely, of a few Dinophyceae.

These algae play a very important role in inland fishery waters. They oxygenate the water during their photosynthesis, they serve as the food of fish either directly or indirectly and, when present in blooms, they present the growth of submerged aquatic weeds, which are a nuisance in most of the fishery waters. It is also likely that some of them, particularly blue-green algae, enrich the water by the fixation of atmospheric nitrogen. The daily photosynthetic activities of the algae are very essential for maintain fish and other animal life in water because they are the only chief primary producer, which can fix solar energy from sun. It is also now known (Schwimmer and Schwimmer, 1955) that several vitamins found in fish can be ultimately traced in the phytoplankton on which they feed.

Maximum number of species were recorded during monsoon and winter season and the minimum during the summer season.

Basing upon the abiotic environmental characteristics and the biotic natures such as an abundance of macrovegetation (*Ramuunculus; Enhydra; Alternenthera*), a predominance of plankton diatoms, the special presence of *Euglena, Daphnia, Cyclops* in summer, and an inhabitation of air-breathing fishes, the shallow pond surveyed can be judged to be one of the "eutrophic" water areas as pointed out by Welch, 1952.

The water depth varied from 0.30 to 0.91m. Due to the high temperature, the water evaporated and the depth of water decreased. With the commencement of high rainfall the depth increased. Water temperature in general showed inverse fluctuation with the depth of the pond. The water temperature varied from 22.2°C to 32.5°C. The water temperature follows closely the changes in the air temperature. Some species of plankton (*Volvox aureus; Closterium acerosum; Ceratium* spp; *Euglena tuba; Simocephalus vetular; Daphnia; lamholari; Diaphanosoma sarssi; D. excisum; Branhiomus quadridentata*) appear in the warmer season.

Dissolved oxygen values ranged from 1.9 to 8.9 ppm during the period of observation. The high dissolved oxygen value obtained in August and the minimum in February can be explained only by photosynthetic activities of the profuse macrophytes, which were distributed uniformly throughout the pond. Nasar and Munshi, (1972) reported the maximum macrophytic biomass in August, 1971 (451.2 g dry wt/m²) in the same pond. In the present study dissolved oxygen values follows closely the changes in the macrophytic biomass. So the amount of dissolved oxygen in this type of ponds depends largely upon the consequent photosynthetic process of floral life of the pond, because during photosynthesis by the aquatic vegetation, assimilation of carbon and liberation of oxygen to the environment take place. The greater abundance of vegetation, the greater the assimilation and consequently the greater the liberation of oxygen.

Dissolved oxygen content showed a wide seasonal fluctuation and this supports the findings of Alikunhi (1957), "When weeds and phytoplankton are abundant the dissolved oxygen in waters undergoes wide fluctuations, sometimes even reaching dangerous limits during night". A similar view has also been expressed by Odum (1957). This might be the reason of discomfort for the fishes which come to the surface and jump out of the water. Only air-breathing fishes can survive under these circumstances. This may be the reason why we do not find the major carps in this pond.

It will be seen from the present findings that the dissolved oxygen content in the pond water does not depend upon the physical factor of solubility. According to the law of solubility of gases, periods of high temperature should be periods of low oxygen content and vice-versa, but that is not so in this case, is evident from the following: (a) From December to February there was no marked change in temperature (22.2°C to 22.7°C) but the oxygen content decreased from 3.8 to 1.9 ppm; (b) from March to April, when the temperature increased from 28.3°C to 32.5°C, the dissolved oxygen also increased from 3.4 to 3.5 ppm; (c) From May to August as the temperature decreased from 30.2°C to 28.3°C, the dissolved oxygen content increased from the above results that the controlling factor of oxygen content in the pond was not the temperature. This supports the findings of Ganapati (1940).

The hydrogen-ion concentration of the pond water expressed in terms of pH was found to vary from a minimum of 7.0 in February to a maximum of 8.5 in July and September. The pH depends upon the amount of carbonates of calcium and magnesium and the carbon dioxide tension in the water. The latter in its turn is influenced by the photosynthetic activities of the aquatic vegetation and the life cycle in the pond.

It is a well known fact that during photosynthesis the carbonates of calcium and magnesium are precipitated from their respective bicarbonates due to the rapid carbon assimilation from the dissolved bicarbonates, and the water becomes more alkaline. The high pH value of the pond water (8.0 to 8.5) would seem to show that the photosynthetic activities are quite high. Atkins and Harris (1924) have taken 8.1 or any value above, to be a sure indication of super-saturation with oxygen. Woodson (1960) remarks on the authority of some phycologist (Hustedt 1939, Fodged 1948, Priscott 1951) that alkaline waters have more species of plants than acidic waters. Hutchinson *et.al* (1929) and Roy (1955) asserted that the high pH is associated with the phytoplankton maxima. These facts may be true in the case of a shallow pond with alkaline water.

Free carbon dioxide was not detected in this pond during the period of investigation. The absence of free carbondioxide can be explained only in two ways; either free carbon dioxide was utilised during photosynthesis by the profuse macrophytes present in the pond as soon as it was formed or the carbonates were found in great (not determined) quantities, so that the free carbondioxide formed during respiration and oxidation of organic matter was not sufficient to convert all of them into bicarbonates. A similar view has also been expressed by Ganapati (1940) in a tank containing a permanent bloom of *Microcytis aeruginasa*. The absence of free $CO_2$ and the abundance of macrophytes, plankton and fishes, show that free $CO_2$ is not a conditioning factor for their production and growth.

Phosphate in the pond soil varied from 8 to 24 ppm, being at the minimum during summer and the maximum during winter season. In this pond, phosphate was found in fairly large quantities throughout the period of the study. Even during the period of intensive photosynthesis phosphate was not entirely absorbed by the profuse enlisted macrophytes. Further, the calcium present in the pond soil plays a very important role at least once a year, after the rainy season, in making the pond rich in nitrogenous matter by the activity of decomposers. This therefore, influences the growth of phytoplankton, which serves as food for zooplankton, and both in turn are available for food of the fishes.

The change in the biological character during different seasons has been correlated with the change in the chemical characters of the pond and with the meteorogical conditions at the pond site. It is observed that winter is not favourable for the growth of macrophytes, and the rainy season seems to be the most favourable season for their growth.

The fishes enlisted are mostly air-breathing, and provided with accessory respiratory organs. Some of the wood fishes such as *Esomus* and *Puntius* are hardy and can survive outside water for quite a long time. All the fishes are carnivores, preying on insects, worms, zooplankton etc. Because of these characteristics, these fishes can withstand adverse condition, such as high temperature, occasional oxygen depletion and also the stagnant condition of the pond. However, they are in great need of these plants for shade, shelter and for breeding purposes. These fishes have indirect food relation with the macrophytes. Almost all the enlisted fishes mainly feed on the live foods which in their turn depend partly on algae or green plants and mainly on decaying animal or vegetable matter. The macrophytes thus support the food organisms of the fish fauna of this pond.

It is further concluded that as unfertilized shallow pond having such characters as alkaline water, without free carbon dioxide, a temperature range of 22.2°C to 32.5°C approximately and profuse macrophytic vegetation is suitable for the inhabitation of air-breathing fishes.

# Swamps and Marshes

Swamps and marshes are wetlands having lentic waters, which form links between terrestrial and aquatic ecosystems. These have been defined by Reid (1961) as shallow water areas with profuse growth of macrophytic vegetation. Beadle and Lind (1960) defined the swamps as sufficiently shallow slow moving water which permit the establishment of characteristic vegetation. Hora and Pillay (1962) and Welch (1952) pointed out that ecologically, the swamps through their different phases of dereliction, are considered to be the final stage of the evolution of land from lakes. Dehadrai and Tripathi (1976) have characterized these water bodies as waterlogged, shallow-water areas with loose peaty bottom, rich in decaying organic matter, retaining water either periodically or shrinking or drying completely during summer months.

The shallowness of water and growth of macrophytic vegetation are undoubtedly the factors for swampy existence. In temperate region, swamps and marshes are often shallow wet depressions, where macro-vegetation is largely of the emergent and floating type expanding over the entire water surface. Technically, a swamp contains persistent standing water along with the vegetation, whereas marshes contain water saturated sediments with no or little standing water among the vegetation.

## Main Types of Swamps

On the basis of extent, magnitude of dereliction and relationship with water sources, Dehadrai and Tripathy (1976) have broadly categorized Indian swamps into two main types.

### Permanent Swamps

These are found where shallow moving or stagnant water remains perennial and protected from wave action and wide changes in seasonal levels. In the regions of North-Bihar (Katihar, Purnea, Samastipur and Begusarai) many such swamps exist. These are larger and smaller tributaries of the complex river systems of Kosi, Bagmati and Gandok in North Bihar with shallow profiles having many slow moving run off channels overgrown with water hyacinths and other floating or submerged vegetation. Besides derelict ditches of various stages in urban and rural areas, usually receiving the common sewage swell at places to form permanent swamps. The bottom silt is very rich in organic matter and other nutrients. A thick coverage of water hyacinth or permanent algal bloom makes these water bodies unsuitable for fishes. Such type of swamps are "doped" or "polluted"

## Seasonal Swamps

These are restricted to land which is only periodically inundated or is subjected to prolonged water logging. The seasonal swamps are of two types: (a) *Senescent swamps*, which are wetted grasslands commonly known as "Chaurs" with a few meter deep water attaining an acute stage of dereliction. These are common in Dharbhanga, Kusheswarasthan, where an area of a few square kilometers is inundated with water during the rainy season either by spilling over of neighbouring rivers, or water received from their catchment areas. Air-breathing teleosts migrate to breed in these shallow waters during the rainy seasons. (b) Semisenescent swamps are shallow water bodies or other depressions usually manipulated for commercial use by aquaculture of cash crops such as 'Makhana' (*Euryale ferox*) and 'Singhara' (*Trapa bispinosa, T. natus*). These activities prevent the water areas from becoming completely swampy and they thus remain in a semi-senescent conditions

## Physico-chemical Factors

Atmospheric temperature and rainfall are the main factors controlling the water level of swamps, especially in the eastern regions of India where significant seasonal rise and fall of temperature and periodic droughts and floods are common occurrences. Besides, the infestation by thick aquatic vegetation involves transpiration leading to heavy evapo-transpirative water loss.

The temperature varies with the concurrent rise and fall of atmospheric temperature. It is well known that the smaller the body of water, the more quickly it reacts to changes in the air temperature. The diel and seasonal temperature variations in swamp is mainly influenced by the amount of sunlight, air temperature, wind velocity and wave action of water.

In general, the swamps have very poor oxygen conditions. The lower layers are often completely deoxygenated. However, a great variety of oxygen conditions exist in Indian swamps. The presence of macrovegetation in water usually induce a hypoxic and hypercarbic aquatic environment, except during period of peak photosynthesis. There exists difference in the levels of dissolved respiratory gases in waters infested with free floating, rooted floating or emergent types with combination of submerged vegetation. Generally dissolved oxygen was lower, dissolved carbondioxide higher, pH lower and temperature fluctuations lower under the water hyacinth areas than in the open water. The extent of differences was found to be dependent upon the time of the day and different seasons (Rai and Munshi, 1979). Due to extreme hypercarbic

**Figure 9.1: Association between Water Hyacinth and Air-breathing Fishes.**

and hypoxic conditions mostly air-breathers and supplemental air-breathers are found living under the coverage of water hyacinth (Figure 9.1).

## Gas Envelops

In order to clearly illustrate these effects, gas envelopes were constructed for different swamps. Such envelopes are represented graphically by plotting all the monthly measurements of dissolved oxygen and free $CO_2$ within a particular swamp and representing these data with a polygon. These envelopes represented the entire set of concentration of dissolved $O_2$ and free $CO_2$ in the studied swamp and pointed out more clearly that the presence of water hyacinth in polluted swamp created a complete hypoxic and hypercarbic water medium distinctly unsuitable for the development of green algae and most of the water breathing animals. More steep envelop in Makhana (*Euryale ferox*) swamps was evident but it was flatter than the gas envelope of *Chaur* swamps which was infested with emergent *Cyperus* reed plants with a combination of sufficient submerged macrophytes. There was comparatively more dissolved oxygen and less free $CO_2$ concentration in *Chaur* swamps than Makhana and DMCH (Darbhanga Medical College Hospital) swamps (Rai, 1980) (Figure 9.2).

Low pH (6.25 – 7.00) of water together with high atmospheric and water temperature, heavy accumulation of organic matter and hypoxic conditions of soils and water were responsible for the accumulation and occurrence of major decomposition products, *viz.* hydrogen sulphide, methane, free ammonia.

The concentration of free $CO_2$ in swamps control the pH and bicarbonate-carbonate alkalinities. Because free $CO_2$ forms carbonic acid ($H_2CO_3$) with water which dissociates into $H^+$ and $HCO^-_3$ ions. This brings a change in the pH of water, as hydrogen ions are set free. $HCO_3$ reacts with calcium or naturally occurring minerals to form bicarbonates, which are soluble. If at this stage free $CO_2$ is not available calcium bicarbonate gets converted into $CaCO_3$. However, if $CO_2$ is in surplus, calcium can be retained in the form of calcium bicarbonate. That amount of $CO_2$ which inhibits

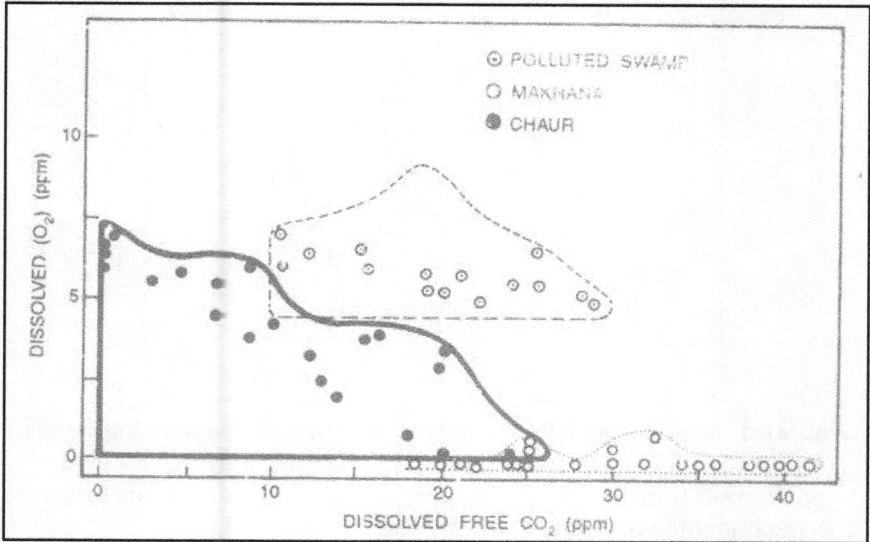

**Figure 9.2: Respiratory Gas Envelope Based on Monthly Variations of Dissolved O$_2$ and Free CO$_2$ Concentrations in different Swamps.**

the conversion of bicarbonates into carbonates is called "equilibrium carbon-dioxide" of Ruttner (1953).

## Nutrient Cycling

The diversification of biotic component of swamps is mainly influenced by nutrient flux. Nitrogen occurs in natural waters in elemental, inorganic and organic forms. The elemental nitrogen in water is derived mostly from the atmospheric and bacterial denitrification of nitrates and ammonia. Phosphate and nitrogenous compounds are usually released into water either by decomposition of organic matters in water and sediments (autochthonous) or by rain, ground and sewage water run-off (allochthonous). These nutrients are utilized and stored by macro-vegetation present in the swamps. A generalized pathway of nutrient movement may be postulated for the exchange of nutrients between plants and sediments of swampy environments:

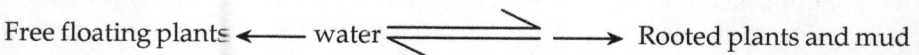

Free floating plants ⟵— water ⇌ ⟶ Rooted plants and mud

## Biotic Component

The biotic component of swampy environments is diverse and varied. These mainly include macrophytes, phytoplankton, zooplankton, periphyton and macroinvertebrates and air–breathing fishes.

## Macrophytes

Four characteristic vegetarians communities were recognized in swamps.These are (i) Free floating type *e.g., Eichhornia crassipes, Trapa bispinosa*, (ii) Bottom rooted floating type *e.g.*, *Euryale ferox*, (iii) Submerged type *e.g.*, *Hydrilla verticillata*, *Potamogeton crispus*, *Najas graminea*, *Ceratophyllum demersum* etc. and (iv) Emergent type *e.g.*, *Cyperus* spp.

There is general agreement that the freshwater swamps have higher macrophytic productivities than any other natural community. Productivity of submerged macrophytes, though considerably high, are always at stressed condition in such habitat, where large floating and emergent aquatic weeds grow. The submerged vegetation was most sensitive to and physico-chemical conditions. Mandal (1988) has shown in Laboratory experiments how much Oxygen (mg/l) is consumed in decomposition of aquatic plants. The problems due to infestation of aquatic weeds are great as they affect the general physico-chemical and biological properties of swamps. However, recently these aquatic weeds are becoming important in the treatment of polluted waters. They absorb potential harmful heavy metals like calcium, nickel, and mercury. The scientists are recommending the use of water hyacinth to remove nutrients from wastewater prior to its discharge into the river.

## Plankton

### Phytoplankton

The phytoplankton communities of swamps are mainly represented by three groups of algae; Cyanophyceae (Blue green algae), Chlorophyceae (Green algae) and Bacillariophyceae (diatoms) (Datta-Munshi, 1993). Usually the phytoplankton were highest during winter and early summer, while lowest during monsoon seasons. Billariphyceae dominated among the phytoplankton communities throughout the year. In North Bihar swamps usually two seasonal peaks of diatoms were found. The primary peak prevailed from winter to early summer while secondary peak started developing in monsoon reaching its maximum in winter.

### Zooplankton

Generally zooplankton communities of swamps belonged to Rotifera, Cladocera and Copepoda. The rotiferan population always dominate over other zooplankton communities in swamps. North Bihar swamps are characterized by rotifer genera like *Brachionus, Filinia, Keratella* and *Polyarthra.*

### Periphyton

The periphyton communities of the swamps mainly include algae and testaceous rhizopods (*e.g.*, *Euglypha, Difflugia, Centropyxis, Arcella* etc.). Usually the periphytic species prefer rich $O_2$, high pH and more alkaline medium. Hard testaceous shells of rhizopods are mainly composed of fine sand particles of silica and decomposed diatom cells. The abundance of Bacillariophyceae and silicate contents of water thus control their numbers in a testaceus rhizopod population.

### Macroinvertebrates

Generally the macroinvertebrates of swamps belonged to Annelida, Gastropoda, Odonata, Ephemeroptera, Diptera, Hemiptera and Coleoptera. The aquatic insects and Gastropods dominate the weed-infested swamps. The shallow standing water with macrophyte infact provide a variety of niches for several insect larvae adapted for boring into the stems and leaves of plants and substrate for benthos. Floating and submerged macrophytes seemed to support larger invertebrate fauna than emergent vegetation.

The fluctuation in the density of macro-invertebrate population depend upon various factors:

i) Stages of life cycle of particular group;

ii) Water depths and vegetation conditions that control directly the and hysic-chemical conditions of water;

iii) Different relationship existing between biotic components in the form of interspecific and intraspecific competition and prey-predator interaction and animal-plant relationships.

Depending upon the degree of associations of macro-invertebrate with aquatic macrophytes, they can easily be differentiated into two major groups; (a) The fauna closely associated with submerged macrophytic parts of vegetation (annelids, chironomids, odonata and ephemeroptera) and (b) others comparatively less associated or generally free moving types (Gastropoda, Hemitera and Coleoptera). These associations depend upon different relationships existing between biology of both plants and animals.

## Air-breathing Fishes

Considerable quantities of fish food material including plankton, periphyton, bottom fauna, insects and decaying organic matter are available in different combinations in swampy environment but their fish fauna is limited. This is mainly due to the adverse physico-chemical conditions of these water so the fast growing, non-predaceous and herbivorous fishes are not able to survive. However, certain slow growing hardy fishes with carnivorous habit and low sensitivity to harmful gases and in hypo-and hypercarbic conditions could survive. The air-breathing teleosts with their predaceous food habits, accessory respiratory organs for aerial respiration and other biological and physiological adaptations can live well in these swamps. Most common air-breathing teleosts inhabiting these habitats are: 4 species of *Channa*, *Anabas testudineus*, *Heteropneustes fossilis*, *Clarias batrachus*, *Monopterus cuchia*, and 3 species of *Mastacembelus*.

## Relationship between Habitat and Air-breathing Habit

The availability of these data should enable interesting ecological conclusions to be drawn about animals inhabiting swamps infested with macrovegetation. With temperature, the dissolved oxygen under macrovegetation depletes rapidly to zero, especially in the summer, when the oxygen demand of most aquatic organisms is greatest. As such most of the vertebrates found in the macrovegetation areas are either entirely dependent upon aerial respiration (snakes, tortoises), or supplemental air-breathers (*Pila*, frogs, air-breathing fishes like the murrels, climbing perches, mud-eel). Further, the dependence upon air-breathing of an organism utilizing bimodal (air and water) gas exchange can be evaluated as a function of the day (Patra *et al.,* 1978) in relation to their stages of life history and their different weight groups.

Two distinct microenvironments in terms of dissolved oxygen and free $CO_2$ were found in the open and vegetation covered water areas. Generally dissolved oxygen was lower, free $CO_2$ higher, pH lower and temperature lower under the water hyacinths than in the "open" water (Rai and Datta Munshi, 1979). A sort of diel fluctuation of oxygen and free $CO_2$ in the two microhabitats have been recorded.

Behavioral studies of these species indicate that they avoid bright light and hide themselves under the coverage of macrovegetation. They come out in the open waters after dusk in search of prey and are more activated at night.

The discovery of circadian rhythm in oxygen consumption in air-breathing fishes is interesting. This is ecophysiological adaptation of air-breathing fishes in relation to the fluctuation of oxygen and carbon dioxide contents of their natural habitat (Patra *et al.,* 1978).

The ability of a murrel (*Channa* spp.) to obtain oxygen from the water will vary with the oxygen tension of the water in swamps and the capacity of the gills to extract the gas from the water. In general the gills are not so well developed in the murrels. All the four species of *Channa* show distinct circadian rhythm in their metabolism. This rhythm seems to be associated precisely with the diel fluctuation of oxygen and carbon dioxide tensions of the water in swamps.

## Conclusion

Thus the two metabolic systems one of habitat (swamp) and other of fishes, are closely interlocked with each other, one influencing the other. The general metabolism of the swamp may be contemplated as a big metabolic wheel, which drives all the small metabolic wheels of different biotic communities. There is some sort of feedback mechanism also in which the metabolic activities of different biotic communities influence the whole metabolism of the ecosystem. Assemblage of airbreathing fishes form an integral part of the swamp ecosystem since their origin several million years ago. The circadian rhythm has now become an inherent property of their system, which they transmit even under the artificial conditions of the laboratory.

# 10

# Diel Variations of Certain Physico-chemical Factors and Plankton Population of a *Chaur* (Wetland) of Kusheswarasthan (Bihar, India)

Certain physico-chemical characteristics, such as temperature, pH, dissolved oxygen, free carbon dioxide, Carbonate/bicarbonate alkalinity, chloride, silicate, phosphate phosphorus and nitrate-nitrogen of surface and 1.5 m deep water showed diel variations with a varied trend. Plankton dominated the surface water during the day. Green and blue-green algae dominated at midday. Maximum density of Zooplankton (584 indiv. $dm^{-3}$) was recorded at 6 h followed by 10 h and 14 h.

Diel variation in the hydrobiological conditions of various freshwater bodies, *e.g.* ponds, lakes, rivers, wells, etc. has been studies extensively (Ganapati 1955, Itazawa 1957, Manny, Hall, 1969, Kant, Kachroo1977, Nassar, 1977, Singh, Saha 1981). Diel variations in the abundance and distribution of plankton of such water bodies in the diel cycle have also been reported (Jarnefelt, 1958, Lorenzen, 1963, Ganf, 1974, Maulood *et.al* 1973). Such biological variations are primarily caused by various physico-chemical characters of the water bodies. Regarding the diel variation in the

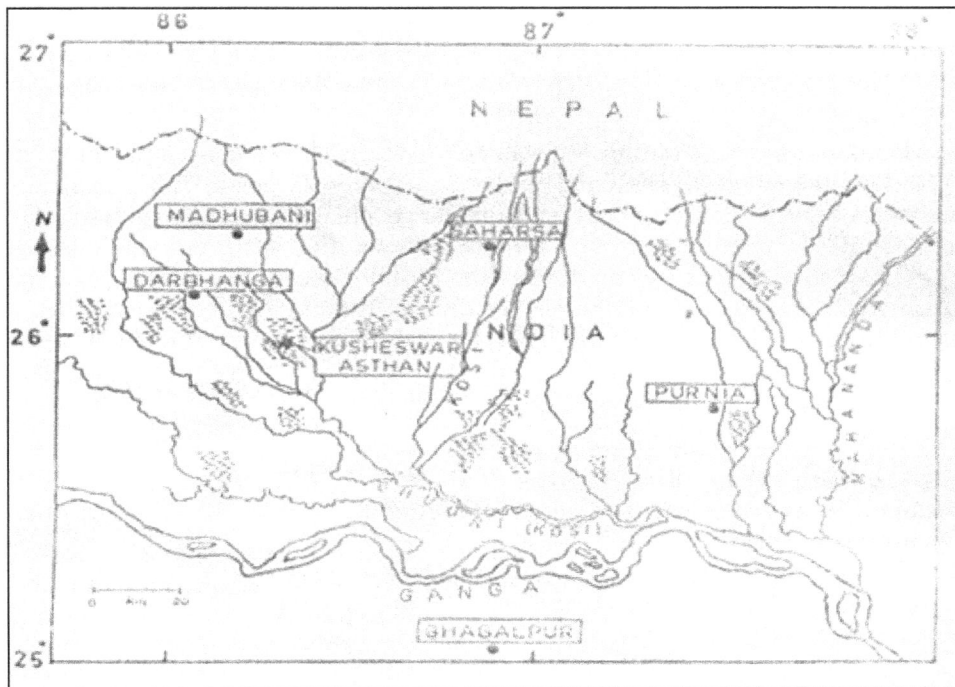

Figure 10.1: Map of the Investigated Area – Location of the Wetland.

hydrobiological conditions of wetlands (chaur), scanty information (Rai, Munshi, 1979) is available.

The aim of the present study was to analyse the diel variation in the hydrobiological conditions and plankton population of a perennial swampy wetland of North Bihar covering about 2 km$^2$ in and around Kusheswarasthan (25° 58′ N latitude and 86° 22′ E longitude, 60 km southeast of the town Darbhanga (Figure 10.1).

## Materials and Methods

Diel variations in some of the physico-chemical parameters and in the plankton population were studied continuously for a period of seven days at four hourly intervals. Four samples from the surface as well as 1.5 m deep for each hour were collected and the mean value was calculated.

Deep water samples were collected with a specially designed apparatus (Figure 10.2). A glass tube was inserted into the water at a particular depth and water sucked in by a suction pump from the first to the second flask. The water in the first flask was free from air contact.

The air and water temperature were recorded by a mercury Celsius thermometer, pH was determined with a Griph pH meter. Dissolved oxygen was determined by

Winkler's methods. Acidity and alkalinity (phenolphthalein and methyl orange) were determined by the method suggested by Welch (1948). Chloride, silicate, phosphate phosphorus and nitrate-nitrogen were determined by standard methods (APHA, 1989).

Plankton samples from surface water only were collected using a plankton net of No.22 bolting silk (75 meshes cm⁻¹) by filtering 100 litres of water. For phytoplankton, immediately after placing the sample into a suitable container, 2 ml of Lugol's solution for each 100 sample volume was added Lugol's solution is a preservative which works as a stain, and causes phytoplankton to sink the quantity of each plankton taxon was calculated as individuals per dm³ by the Lackey drop microtransect counting method (1938). The data were statistically analysed by Student's "t" test.

# Results

## Physico-chemical Condition of Water

The mean values of the data based on seven days observations on the diel variation in hydrological parameters of wetland water of "Kusheswarasthan Chaurs"are given in Table 10.1.

The air temperature varied from 23.5 to 37.0°C with a maximum at 14 h and minimum at 2 h. The water temperature fluctuated from 26 - 35°C having no fixed diel variation range. It had its maximum during the day (14 h) and minimum at midnight and early morning (2-6h).

The pH was acidic throughout the diel cycle except at 10 h when it was slightly alkaline. Its minimum (pH=5.0) was at 18 h and maximum (pH = 7.2) at 10 h both in surface and deep water.

Dissolved oxygen (DO₂) showed marked variation in both surface and deep water in the diel cycle. In surfacing water it varied from 0.2-6.2 mg dm³ and in the deep water from 0.2 (at 2 h) to 1.3 mg dm³ (at 14 h). The oxygen values showed a gradual decrease from 10 to 2 h except at 22 h where there was a slight deviation. However, a similar trend was observed after 14 h in deep water. The concentration of DO₂ in surface and deep water showed a very highly significant difference at 10 h (Table 10.2) (t-118.83; P<0.001).

Free carbon dioxide (FCO₂) in surface water ranged from 0-12.4 mg dm³. In deep water it was absent at 14 h and had a maximum at 10 h when a very significant difference occurred between the surface and deep water (t=49.59; P<0.001).

Carbonate alkalinity was absent in almost all diel cycles except at 10 h in surface and 14 h both surface and deep water. A very significant difference between surface and deep water was also shown (t=73.44; P<0.001) at 10h.

Bicarbonate alkalinity was recorded in both surface and deep water throughout the diel cycle. In surface water its value ranged from 102-128 mg dm⁻³ and in deep water from 112-142 mg dm⁻³. In surface water it reached a maximum at 18 h and a minimum at 14 h, whereas in deep water the maximum and minimum values were recorded at 6 h and 22 h respectively. The bicarbonate alkalinity of surface and deep water showed the maximum significant difference (t=8.51; P<0.001) at 6 h.

**Table 10.1: t-values for Mean difference of Chemical Characters at Surface and in Deep Water at Various Study Hours.**

| Hours | $DO_2$ | $FCO_2$ | $CO_3^{2-}$ | $HCO_3^-$ | Chloride | Silicate | $PO_4$-P | $NO_3$-N |
|---|---|---|---|---|---|---|---|---|
| 06.00 | 1.01[ns] | 5.56** | 0 | 8.51 | 102.52 | 39.54 | 13.66 | 11.10 |
| 10.00 | 118.83 | 49.59 | 73.44 | 0 | 4.01** | 10.30 | 1.33[ns] | 35.85 |
| 14.00 | – | 0 | 13.89 | 3.92** | 4.30** | 6.55 | 15.68 | 321.86 |
| 18.00 | 12.09 | 12.44 | 0 | 4.08** | 9.58 | 24.15 | 7.00 | 29.16 |
| 22.00 | – | 26.92 | 0 | 1.46[ns] | 20.44 | 52.47 | – | 6.70 |
| 02.00 | 2.37[ns] | 34.71 | 0 | 0.88[ns] | 3.02* | 1.04[ns] | 3.27* | 10.47 |
|  | –7.95 |  |  |  |  |  | 23.49 |  |
|  | 4.55** |  |  |  |  |  |  |  |

Ns: Not significant;

*: p<0.65; **: p < 0.01; 0: Shows no difference.

All others values are highly significant at 0.1 per cent level (p<001)

Chloride concentration was higher in deep water than in surface water. In the latter it reached a maximum (17.6 mg dm$^{-3}$) at 2 h and minimum (12.4 mg dm$^{-3}$) from 14-22 hours. In deep water maximum (32.9 mg dm$^{-3}$) and minimum (13.4 mg dm$^{-3}$) values were recorded at 6 h and 14 h respectively. The maximum significant difference between surface and deep water was recorded at 6 h (t=102.52; P<0.001).

Silicate concentration reached a maximum (14.3 mg dm$^{-3}$) at 14 h and minimum (6.2 mg dm$^{-3}$) at 6 h in surface water, while in the deep layer it was maximum (25.0 mg dm$^{-3)}$ at 22 h and minimum (6.7 mg dm$^{-3)}$ at 6 h. Surface and deep water showed highly significant differences except at 2 h.

In deep water, phosphate phosphorus was absent in almost all diel cycles except at 6 h (0.03 mg dm$^{-3)}$ and 10 h (0.16 mg dm$^{-3}$). In surface water its value ranged from 0.05-0.33 mg dm$^{-3}$. It reached a maximum at 14 h and minimum at 22 h (t = 11.76; P<0.001). Both surface and deep water showed significant variation in the diel cycle except at 10 h.

$NO_3$-N showed marked diel variation in both surface and deep water, the latter showing higher values only at 14 to 18 h. In surface water, it was maximum (3.20 mg dm$^{-3}$) at 10 h and minimum (0.55 mg dm$^{-3}$) at 6 h. Deep water showed a high maximum concentration (7.20 mg dm$^{-3}$) at 14 h when the difference from the surface water was statistically highly significant (t=321.86; P<0.001).

## Vertical Migration of Plankton

Phytoplankton density showed a clear diel variation. The members of Myxophyceae, Chlorophyceae and Bacillariophyceae, however, did not show the same trend of diel variation. Blue-green algae showed maximum density (983 indiv. dm$^{-3}$) at 14 h and minimum 73 indiv. dm$^{-3}$ at 2 h. Green algae showed a remarkable preference for bright day light, with a maximum (935 indiv. dm$^{-3}$) at 10 h, 14 h ranking second with a density of 704 indiv. dm$^{-3.}$ At midnight it was reduced to

51 indiv. dm$^{-3}$). Minimum density (62 indiv. Dm$^{-3}$) was recorded at 22 h. The phytoplankton density showed an increasing trend from a minimum at 2 h to a maximum at 10 h, thereafter having a decreasing trend.

Zooplankton showed more distinct diel variation than phytoplankton in the surface water. Maximum total density (584 indiv. Dm$^{-3}$) was recorded at 6 h and minimum (164 indiv. Dm$^{-3}$) at 22 h. Ciliates of Protoza showed dominance at 6 h with 220 indiv. Dm$^{-3}$ then followed a decreasing trend with the increasing diel cycle and their minimum (51 indiv. Dm$^{-3}$) was noted at 22 h. Again at 2 h, the density was somewhat higher. Rotifers reached their maximum density at 10 h (112 indiv. Dm$^{-3}$), showing a decreasing trend until 2 h. The trends for copepods were similar to those of protozoans. Adult copepods were found in maximum density (119 indiv. Dm$^{-3}$) at 6 h while they were absent at 22 h. Copepods start an upward movement at midnight and reach a maximum in the morning hours near the surface. However, after sunrise a downward movement began and at night (22 h) its population was absent from the surface. Cladocerans showed a somewhat similar pattern as protozoans and copepods. Their maximum density (153 indiv. Dm$^{-3}$) was recorded at 6 h and minimum (42 indiv. Dm$^{-3}$) at 14 h. A gradual increase in cladoceran population was recorded after 14 h until reaching a maximum at 6 h.

The descriptive model shows that there was a relationship between temperature and several chemical characters of the swamp water. The water temperature was higher during the day when other chemical factors were either higher or lower, indicating that temperature had a marked influence on these factors. The density of phytoplankton (Figure 10.4) was greater in surface water during the day 6-18 h), when several chemical factors, *i.e.* pH, DO$_2$, CO$_3^{2-}$, HCO$_3$, chloride, silicate, PO$_4$-P and NO$_3$-N were higher.

FCO$_2$ was at its maximum during the night owing to the decreased photosynthetic activity of phytoplankton. Similarly as the surface water, the deep water also showed that several chemical factors were higher during the day. Among these factors only the higher FCO$_2$ during the day in deep water is indicative of a non-photic nature of the zone.

A higher oxygen content was recorded during the day in both surface and deep water, which may be attributed to the increased photosynthetic activity of phytoplankton. The differences in the oxygen content of both surface and deep water at different hours were highly significant ($t = 162.67$; $P<0.001$ at 2 and 10 h in surface water; $t = 30.96$; $P<0.001$ at 2 and 14 h in deep water). In both zones at night the oxygen content was depleted, owing to respiration of the aquatic organisms. During the night (22 to 2 h) there was no photosynthetic activity, but even then the surface water column maintains its oxygen content as surface water receives O$_2$ from the air and possibly the plankton migrate to deeper water. In contrast, the deep water showed a lower oxygen content than the surface water because of the increased planktonic population. The salinity was at a minimum during the day and a maximum at night. The differences were highly significant ($t = 17.68$; $P<0.001$) in surface water at 14 and 2 h. The high salinity value may be due to the rapid evaporation, large surface area, and bathing activities. Again the deep water showed a higher chloride content than the surface water, which might be due to the deposition of chloride from the surface to deep water.

This work provides evidence that the timing and amplitude of physical, chemical and biological changes occurring in a swamp are related to the diel periodicity of environmental variables. The diel responses of the phytoplankton to such events may be due to their direct responses to environmental variation during the diel cycle (exogenous rhythms) and entrained responses (endogenous rhythms). Among phytoplankton, members of Chlorophyceae and Myxophyceae showed maximum density at 6 to 14 h and minimum at 22h. This may be attributed to an upward movement of the phytoplankton in the early morning, climaxing at noon, followed by a downward migration.

Grazing by zooplankton may also affect the diel distribution of phytoplankton in the water column. The affinity of the zooplankton towards the phytoplankton population may be attributed to their upward migration in search of food. As zooplankton feed upon phytoplankton a negative correlation is to be expected at the surface water. However, there is a positive correlation between phyto-and zooplankton (t=2.934; P<0.05). This seems to be due to the over production and high turnover rate of the phytoplankton. Again, light and physico-chemical conditions of the surface water are conducive to the production of phytoplankton. Among zooplankton, cladocerans utilize phytoplankton as their food more than do other members of zooplankton because of their larger size. Cladocerans showed a preference for diatoms (r=0.755).

The density of zooplankton was at a maximum at 6 h while there was a distinct decreasing trend with the passing hours until night (22 h). During diel migration zoo and phytoplankton are linked together owing to their respective positions in the food chain and specific type of environment.

# 11

# Biological Productivity And Energetics

The biological productivity of any system whether aquatic or terrestrial involves the trapping of solar energy by chlorophyll bearing plants and its transformation within the system by different organisms at different trophic levels. The energy that enters the earth's surface as light is balanced by the energy that leaves the earth's surface as invisible radiation. The rate at which energy is stored by green plants is called primary productivity and by the heterotrophs secondary productivity. The essence of life is the progression of such changes as growth, self-duplication and synthesis of complex relationships of matter which actually depends upon the biological productivity and energetics within the system.

## Concept of Productivity

There are four successive steps in the process of biological productivity:

### Gross Primary Productivity

It is the total rate of photosynthesis including the organic matter used up in respiration during the measurement period. This is also known as total photosynthesis or total assimilation.

### Net Primary Productivity

It is the rate of storage of organic matter in plant tissues in excess of the respiratory use by the plants during the measurement period. This is also called apparent photosynthesis or net assimilation.

## Net Community Productivity

It is the rate of storage of organic matter not used by heterotrophs during the growing season or a year. Actually this is the net primary production minus the heterotrophic consumption during the period of consideration.

## Secondary Productivity

It is the rate of energy storage at consumer level. The consumers utilize food materials produced with the respiratory loss at authotrophic level and convert them to different tissues by the overall process of assimilation. The secondary productivity should not be divided into gross and net amounts. The increase in body weight after loss of organic matter and energy in the form of body heat and respiration at heterotrophic level is the secondary productivity.

# Primary Productivity in Freshwaters

The primary productivity of an ecosystem is defined as the rate at which radiant energy is converted by photosynthetic and chemosynthetic activity by producer organism to organic substances. Though primary productivity is the product of photosynthesis, yet these two processes are not identical. The uptake and incorporation of inorganic nutrients in the presence of water contribute towards the primary productivity. Temperature governs annual productivity in various ways.

The freshwater of the earth comprises less than 4 per cent or the total earth surface. Most of the surface of freshwater exists as ice and snow in glaciers and polar ice caps (approximately $25.5 \times 10^6 km^3$). Freshwater lakes and streams cover about 0.2 per cent and have an estimated area of about $2 \times 10^6$ km$^2$. The mean net productivity is 200 gC/m$^2$/year. Freshwater marshes and swamps comprise almost the same area of about $2 \times 10^6 km^2$, but give a comparatively much higher mean net productivity value 1500 gC/m$^2$/year.

# Carbon Fixation

Photosynthetic fixation of carbon in freshwaters may occur by various communities. These communities may be phytoplankton, macrophytes or periphyton. The phytoplankton represents the algal community of the open water. The macrophytes are macroscopic vascular plants which are either submerged or emergent, rooted or floating. The periphyton grow on submerged substances. In many cases diatoms are the dominant periphyton forming a film on the surface of mud, rocks, sands or on the surfaces of macrophytes. In Table 7.1 the net primary productivity by different aquatic communities are given. The relative importance of these three groups of primary producers in aquatic systems is highly variable. In very large and deep lakes phytoplankton is the major primary producer. However, in the shallow shore line areas of lakes, rivers, swamps and marshes the contribution by macrophytes and periphyton in carbonification process is certainly very high. (Table 11.1).

The photosynthetic bacteria though play a minor role in the production of organic matter can function under unfavourable conditions. The green and purple sulphur bacteria which are obligate anaerobes occur in the boundary layer between oxidized and reduced zones in sediments or water where there is low insencity of light. Photosynthetic sulphur bacteria accounts for 3 to 5 per cent of the total annual production in most lakes. In stagnant lakes rich in $H_2S$ these bacteria may contribute up to 25 per cent of the total production. Though bacterial photosynthesis is of considerable importance in polluted and eutrophic waters, it is no substitute for the oxygen generating photosynthesis on which the biosphere depends.

**Table 11.1: Net Primary Productivity Values for Aquatic Communities (Leith and Whittaker 1975)**

| Producer Community | Mg C/m²/day | G Dry Organic Matter/m²/year |
|---|---|---|
| Lake phytoplankton Freshwater macrophytes | 100 – 1200 | 100 – 900 |
| Submerged | 500-2700 | 400 – 2000 |
| Emergent | 4100-12000 | 3000 – 8500 |

The utilization of ceterrent carbon, whether autochthonous (originating within the system) or allochthonous (originating outside the system) may be important in the total metabolic system of lakes and streams. Allochthonous carbon inputs also can contribute significantly to the total metabolism of aquatic ecosystem through the same pathways of grazing and decomposition.

## Measurements of Primary Production

The units of measurements of primary production are often confusing and difficult to compare, e.g., milligrams dry weight, milligrams of ash-free dry weight, milligrams of glucose milligrams carbon, milligrams $O_2$, millimoles $O_2$ or $CO_2$, kilocalories etc.

There are many techniques for their measurement. Such methods as harvest, gas exchange, change in pH, radio iosotopic tracer and other methods have been used for the estimation of primary productivity, $C^{14}$ method is widely used for measurement of carbon fixation. But in this method there is artificiality of environment, with effects of incubation time and conditions, effects of light and temperature, shock during sample manipulations, and formation of extracellular product. The accuracy of measurements of productivity of an aquatic system depends on varied factors like depth of trophozenic and mixing within the water column etc. Thus there are often major temporal (annual, seasonal, and daily) and spatial (vertical and horizontal) differences in the rates of primary productivity within an aquatic ecosystem. If careful studies are made with frequent measurements in all the seasons a reasonably good estimate of annual primary productivity is possible.

## Magnitude of Production

Westlake (1966) suggested that annual net primary productivity is roughly 50 per cent of gross productivity in aquatic system. The aquatic macrophytes fix 40 per cent to 48 per cent carbon. Likens (1975) approximated value for respiration as 40 per cent of gross productivity.

Rivers range down to a minimum of < 1 mg carbon fixed per meter square per day because most of the rivers and streams have very little autochthonous production and depend mostly on allochthonous inputs. There is difficulty in studying the primary productivity of these flowing water systems. In fact a parcel of water at one location in a river is constantly changing as it moves downstream.

Arctic, Antarctic and alpine lakes are less productive because of limitation in the growing season (covered by ice and snow during part of the year) and because of the seasonal delay or limitation in recharge of nutrients from the drainage system.

Rodhe (1969) summarized data for Danish lakes and Swedish lakes, all fertilized by domestic or industrial effluents which gave a maximum daily rate of 6000 mg C/m² for productivity. Lake Baikal which has largest volume of freshwater in the world, has an average phytoplankton productivity of about 310 mg C/m²/day (Kozhov,1963), an annual net productivity of 122.5 g C/m² (Moskalenko, 1972) and is probably mesotrophic (Table 11.2).

**Table 11.2: Net Primary Productivity in Aquatic Ecosystems (from Likens 1975)**

| Freshwater Systems | mg C/m²/day | gC/m²/year |
|---|---|---|
| Tropical lakes | 130-7600 | 30-2500 |
| Temperate lakes | 3-3600 | 2-950 |
| Ancient lakes | 1-170 | <1-35 |
| Antarctic lakes | 1-35 | 1-1 |
| Alpine lakes | 1-450 | <1-100 |
| Temperate streams | <1-3000 | <1-650 |
| Tropical streams | 1-1000 | 1-1000 |
| Oligotrophic lakes | 50-300 | – |
| Mesotrophic lakes | 250-1000 | – |
| Eutrophic lakes | 600-8000 | – |
| Dystrophic lakes | <50-500 | – |
| Ganga river | – | 592* |

* Bilgrami and Munshi 1982-85.

Tropical lakes may be exceedingly productive if nutrients are plentiful. It has been known for a long time that very high productivities may be obtained in shallow tropical ponds and paddy fields. For the upper limit of the aquatic primary productivity in natural ecosystems, Wetzel (1966) suggested that only in highly enriched situations it is likely to exceed 5 g C/m²/day.

In India the primary productivity of various freshwater systems (Table 11.3) ranges from very low values of 90-100 g C/m²/year in Kashmir Himalayan lakes to as high as 1130 to 3309 g C/m²/year in sewage doped eutrophicated swamps and Makhana swamps of North Bihar. Even the side spill over channel (Champanala) of the river Ganges is quite productive (876 g C/m²/year). The lakes with low fertility level had a phytoplankton dominance of diatoms and chlorophyceae, whereas in eutrophic lakes *cyanophyceae* predominate.

## Secondary Productivity

Secondary productivity is the rate of formation of new organic matter by heterotroph. The secondary production depends upon the amount of primary production as well as the quality and the diversity of the producers.

#### Table 11.3: Net Primary Productivity in Freshwater Lakes, Ponds and Streams of India

| Freshwater System | Net Primary Productivity g C/m²/Year | Authors |
|---|---|---|
| Southern India Ooty lake | 1688 | |
| Sandynalla lake | 864 | Ganapati (1972) |
| Kodai lake | 110 | |
| Yereand lake | 1058 | |
| Kashmir Himalayas | | |
| Nilnag lake | 90-100 | Khan and Zutshi (1980) |
| Bihar Bhagalpur pond | 1095 | Nasar and Datta Munshi (1975) |
| Champanala side Spill over channels of Ganga river | 785 (Phytoplankton) 91(Macrophytes)--876 | Husan (1982) |
| Badua reservoir | 1314 | Verma and Datta Munshi (1983) |
| Ganga river | 592 | Bilgrami and Datta Munshi (1982-85) |
| Sewage doped swamps | 156 (Phytoplankton) 974 (Macrophytes)--1030 | Datta Munshi and Datta Munshi (1985) |
| Makhana swamps | 172 (Phytoplankton) 3137 (Macrophytes)--3309 | Datta Munshi and Datta Munshi (1985) |

## Producer-consumer Relationship

The primary consumers (herbivores) feed directly on living plants or plant remains. In aquatic systems they are zooplankton and benthos (bottom forms). The secondary consumers (carnivores) such as predaceous insects and fishes feed on primary consumers. Some of the consumers consume plants as well as animals and are known as omnivores. Another important type of consumer is the detritivore which depends on organic detritus for their nourishment.

In freshwaters it is usually possible to distinguish two trophic levels, - the pelagic (open water) trophic levels and benthic (bottom) with their distinctive producer consumer food web. The pelagic producer-consumer chain begins with the phytoplankton, followed by the herbivorous group of zooplankton and then by the predatory ones. Both the groups of zooplankton are eaten by pelagic fishes. The predatory fishes are the terminal consumers.

The benthic producer-consumer chain starts with benthic plants like *Chara,Potomageton* etc. with sedimented organic and other allochthonous organic matter forming detritus and leads via bottom-dwelling animals like Tubifex and Chironomus to fish species like *Labeo calbasu* which derive their nourishments largely from the floor of the lake or streams. However, these two consumer chains are inter related with each other and the benthic profundal fauna of lakes and deep stream's depend on organic matter produced in the pelagic zone (Figure 11.1 a-n).

## Assimilation of Food

It is important to know as to what extent the producers are available for utilization by the consumers and what fraction of the consumer's diet is assimilated and passed on to the next stage in the food chain.

With the help of the following formula (Iviev 1939; Winberg,1956) the relationship between total food intake and assimilation of food by the consumers may be explained.

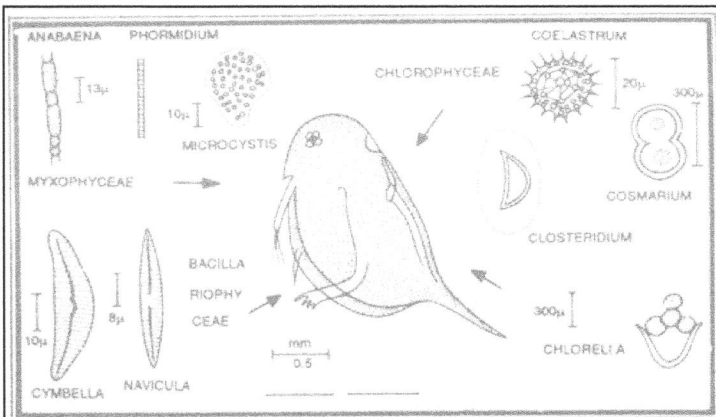

**Figure 11.1a: Food Items of *Daphnia carinata.***

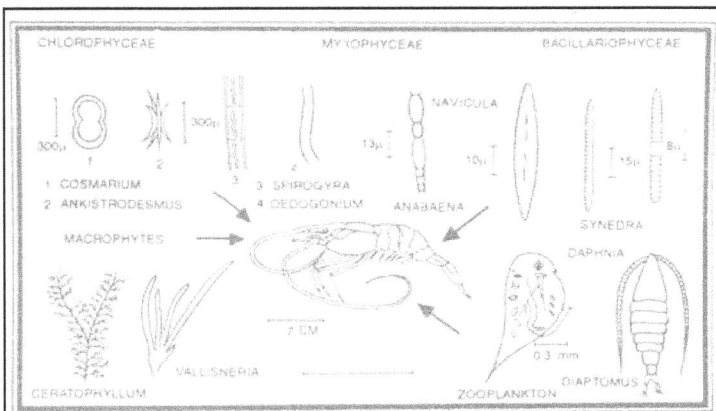

**Figure 11.1b: Food Items of *Macrobrachium.***

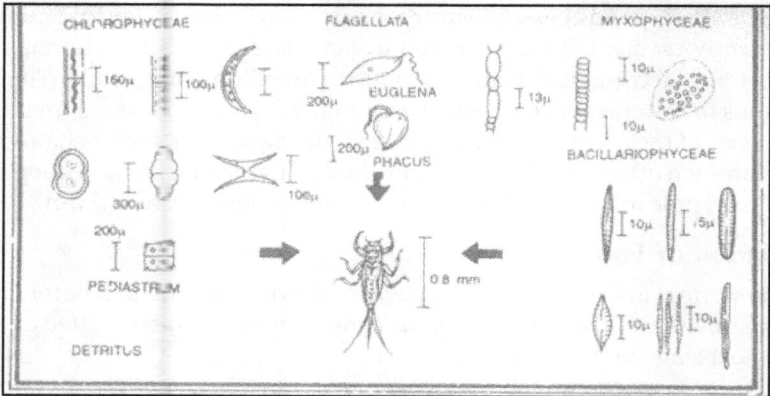

**Figure 11.1c: Food Items of *Ephemerella* (Nymph).**

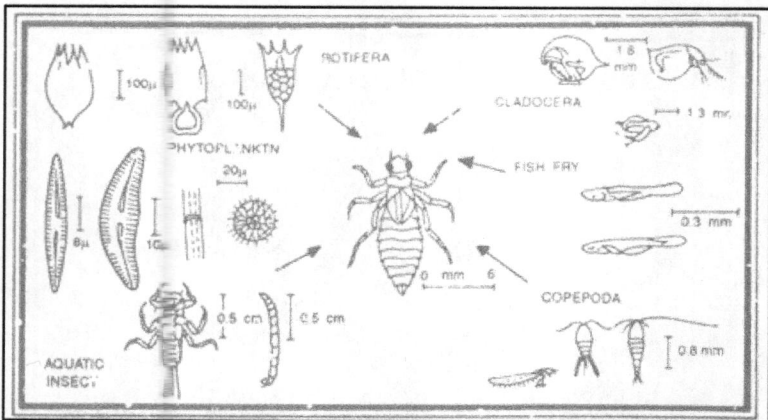

**Figure 11.1d: Food Items of *Mesopomphua* (Nymph).**

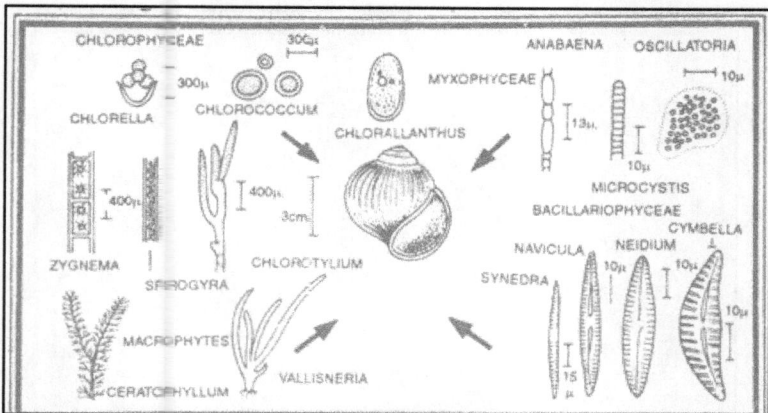

**Figure 11.1e: Food Items of *Pila globosa*.**

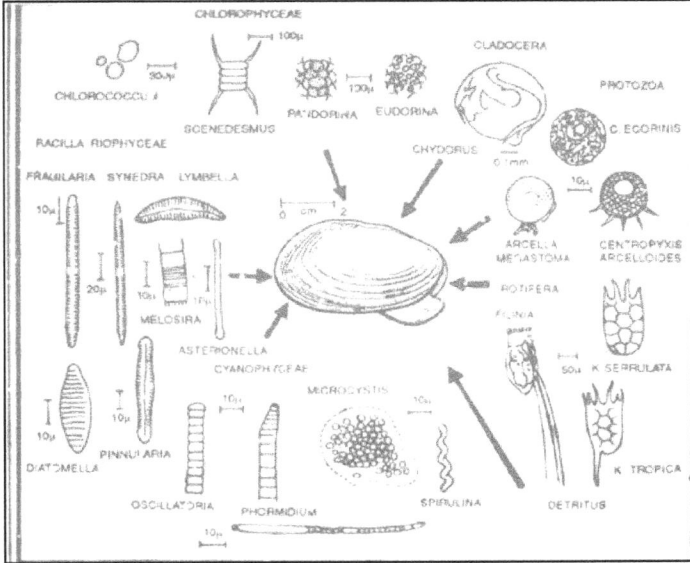

**Figure 11.1f: Food Items of *Parreysia favidens*.**

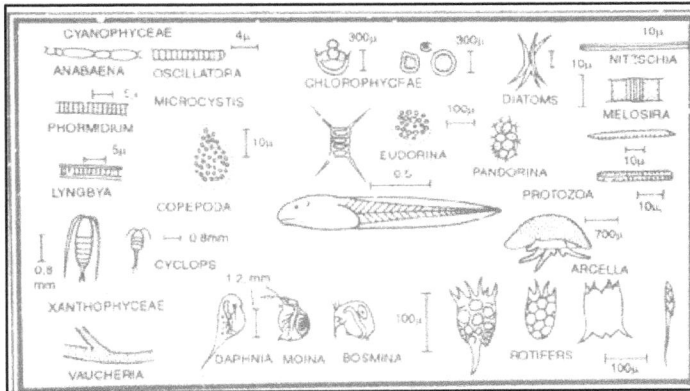

**Figure 11.1g: Food Items of Tadpole Larvae.**

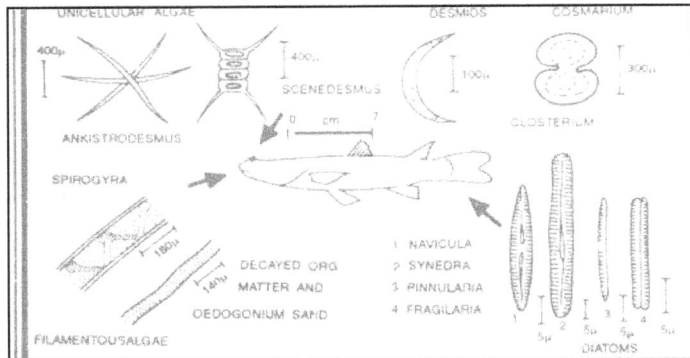

**Figure 11.1h: Food Items of *Rhinomugil corsuls*.**

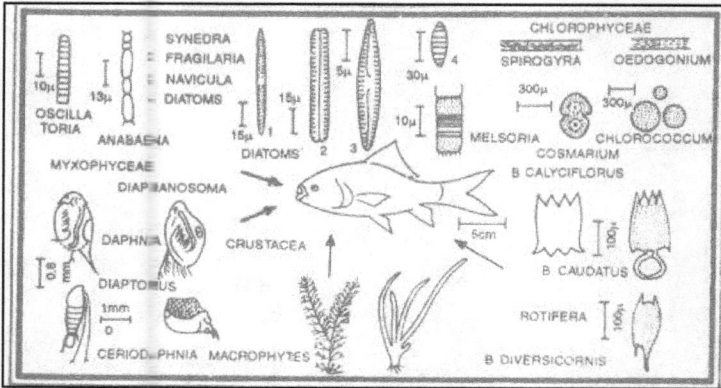

**Figure 11.1i: Food Items of** *Catla catla.*

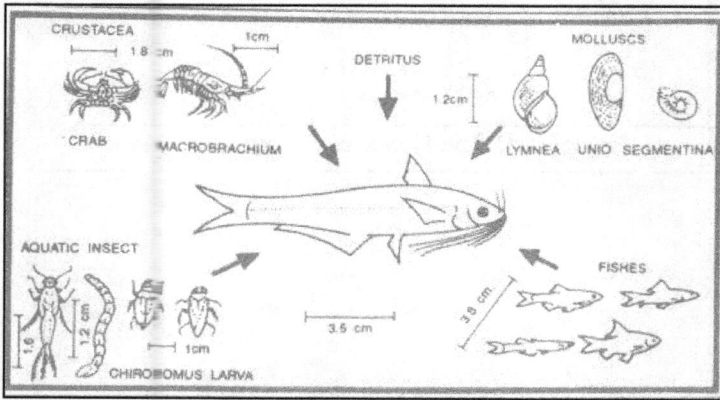

**Figure 11.1j: Food Items of** *Clupisoma garua.*

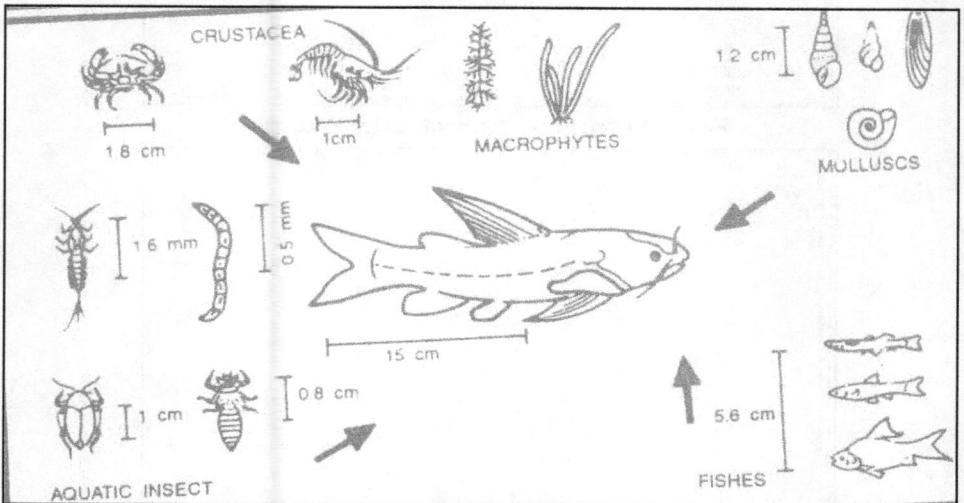

**Figure 11.1k: Food Items of** *Rita rita.*

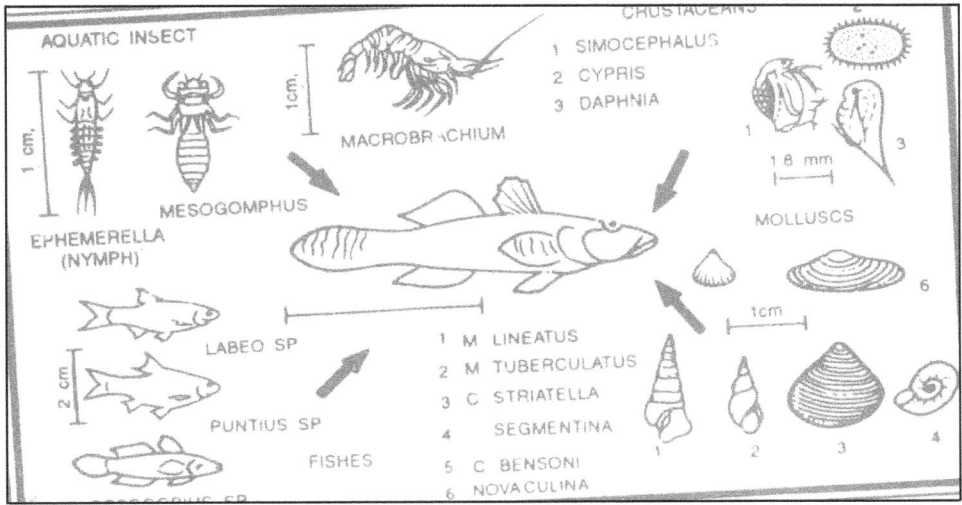

**Figure 11.1l: Food Items of *Glossogobius giuris*.**

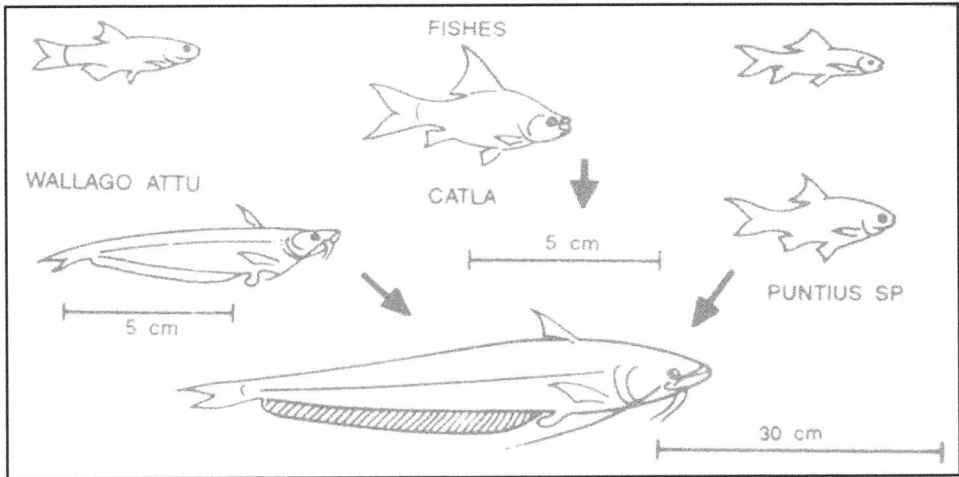

**Figure 11.1m: Food Items of *Wallago attu*.**

$$I = G + R + E$$

From the total food intake (I), a part will be used to supply energy for respiration and other metabolic purposes, is part incorporated into body tissues for growth (G), and the remainder is excreted unused (E). The amount of assimilated diet (A) in the form of energy is equal to the rate of intake (I) minus the rate of egestion (E):

$$A = I-E \text{ and hence } A = R+G$$

Thus Assimilation (A) at consumer's level is compared with Gross production at producer level, while Growth (G) or increase in body weight of consumers is

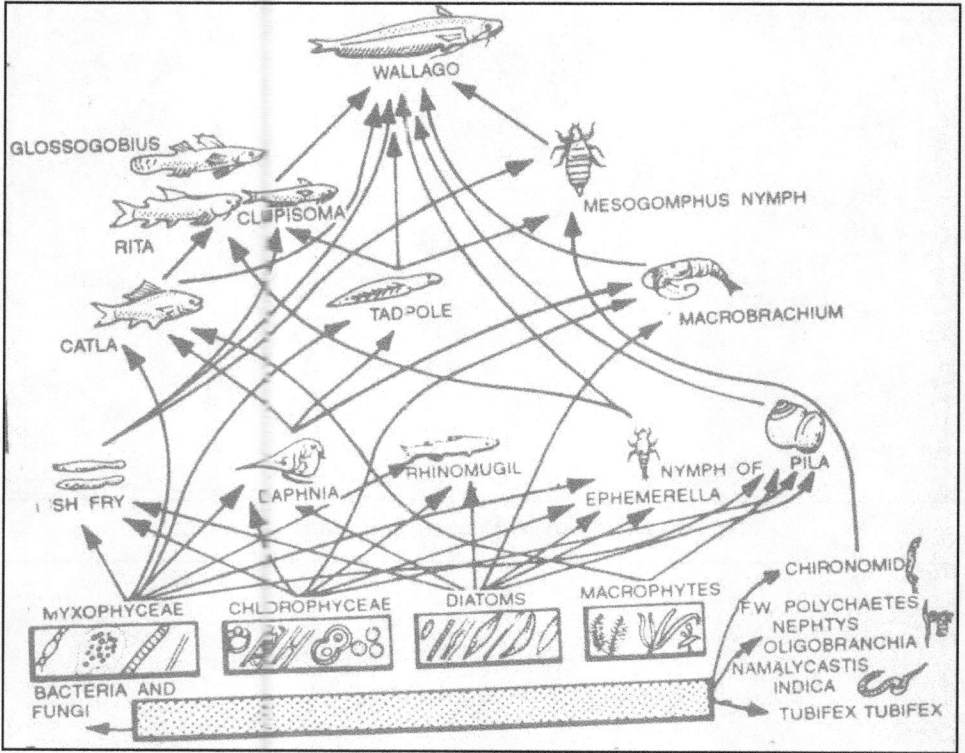

**Figure 11.1n: Trophic Relationships of Biota and the Formation of Food-Web in Ganga River at Kahalgaon.**

equivalent to net production of producers and is denoted as secondary production (P): G = P at consumer's level. Some metabolic efficiencies are important for ecological studies at consumer's level:

A/I = Assimilation efficiency

G/I = Gross growth efficiency, $K_1$

G/A = Net growth efficiency, $K_2$

## Secondary Productivity

The secondary production has been defined as the increase in biomass which occurs in a given period of time. Many terms are in use.

The secondary productivity (P) may be at primary or secondary consumer's level. When the yolk sac stage of a fish is passed, it begins to capture prey, mainly zooplankton. As they grow in size they change their food habits. The carnivorous fishes become predator as they grow in size. The distinction between secondary or tertiary consumers at the juvenile stage of fishes is uncertain.

The relationship between the density of zooplankton and energy spent in the capture of prey was studied on the bream fish (*Abramis brama*) fed with radioactively-labelled Bosmina sp. (a cladoceran of water). At a field density of 30 zooplankton per litre, the young fish used up more energy in the process of capture. At a density of 45 to 60 per litre the energy balance was equalized. At a density 60 to 80 per litre there are positive balance of energy and the fish began to grow and the growth was optimized at 500 to 800 organisms per/litre (Sorokin and Panov, 1966).

For fisheries management in freshwater and for the cultivation of fish in ponds the population of fish must be regulated to the amount of food available in such a way that the fish achieve their maximum growth. As such the primary productivity in managed fish farms must be increased first with fertilizer treatments. This enhanced primary productivity will increase the population density of the feed organisms on which the fish populations depend.

## Standing Stock

It is defined as the amount present at a point in time, expressed best as quantity over spatial unit. Synonyms include standing crop, stock and biomass. Units employed for standing stock include weight (usually either in grams or kilograms), either wet or dry weight or ash-free dry weight, carbon and calories per m$^2$ or ha and expressed as g.m.$^{-2}$ (dry), kg/ha (wet/dry), kcal/m$^2$.

## Production or Productivity

Is defined as the rate of tissue elaboration, regardless whether it survives to the end of a given period of time. The units employed are same as for standing stock with the addition of the time unit and expressed as kg.ha$^{-1}$.yr$^{-1}$(dry weight), g.m$^{-2}$.yr$^{-1}$, kcal, m$^{-2}$.yr$^{-1}$(calorific value). There have been very few studies of secondary production at trophic level and most of the works relate to production by individual species.

## Productivity of Molluscs

As far as the productivity of mollusks are concerned very little literature is available. Negus (1966) determined the growth rate of freshwater mussels *Anodonta* and *Unio* from the winter rings of their shells. Paine (1971) worked on the seasonal changes in population dynamics of the intertidal gastropod *Tegula funebralis*. Hunter (1975) studied the linear and biomass growth and bioenergetics of three populations of *Lymnaea palustris*. Similarly, Burky (1974), Browne (1978) and Aldridge *et. al.* (1978) worked on growth, fecundity, biomass and productivity of different freshwater snails. Prieto *et.al* (1985) studied the secondary productivity of the mussel *Modiolus modiolus*, James (1985) studied the distribution, biomass and production of the freshwater mussel *Hyridella menziesi*. Craeymeersch *et. al.* (1986) studied the secondary production of an intertidal mussel *Mytilus edulis* L. In India, the literature and work related to secondary productivity of mollusks are those of Haniffa and Pandian (1974,1978), Vivekanandan *et. al* (1974) and Haniffa (1978a,b).

Singh (1988) studied secondary productivity of the common prosobranch of Bihar, *Pila globosa* and *Bellamya bengalensis* in fixed areas of a tank from August, 1985 to July, 1986. The samples were collected using a wooden frame (30x30 cm) and by hand picking method. The samples were brought in polythene bags to the laboratory, washed, counted and weighed and were dried at 65°C until constancy for dry weight was obtained.

Secondary productivity of the prosobranchs were determined by the dry weight method and expressed as g. dry wt.m$^2$.yr$^{-1}$.

Population density of *Pila globosa* ranged between 10.48 m$^{-2}$ in September/October to 54.45 m$^{-2}$ in March/April. Young ones were abundant during November/December, intermediate ones from January to May and old ones from February to April. December to May represents the time of maximum abundance of *P.globosa* in Monghyr, Bihar. Biomass ranged between 35.86 g.m$^{-2}$ (dry weight) in October to 153.54 g.m$^{-2}$ in April. It increases from October onward till April, and then starts decreasing. Mean biomass for the whole year was 97.37 g.m$^{-2}$ and mean density of *P. globosa* was 26.85 g.m$^{-2}$.

The secondary productivity of *Pila globosa* was 117.68 g. m$^{-2}$ yr$^{-1}$ *i.e.* 326.9 mg. m$^{-2}$.day$^{-1}$. On the other hand the population density of *Bellamya bengalensis* varied from 198.0 m$^{-2}$ in August and mean density was 381.76 m$^{-2}$. Biomass ranged between 52.37 g.m$^{-2}$. Secondary productivity of *B. bengalensis* was 146.86 g.m$^{-2}$ yr$^{-1}$, i.g., 399.6 mg.m$^{-2}$.day$^{-1}$.

## Biomass and Production of Aquatic Insects

Roy (1986) has investigated biomass and productivity of aquatic insects in a freshwater pond. The maximum production of the aquatic insects population was found in summer and monsoon seasons when positive increase in biomass was recorded. The maximum production rate was recorded in November (0.028 g cubic meter$^{-1}$.day$^{-1}$) and minimum (0.003 g cubic meter$^{-1}$.day$^{-1}$) in January. The annual productivity comes to 5.905 g. cubic meter$^{-1}$ month$^{-1}$ (Table 11.4). It was found that the production of these insects in this pond is governed by complex extrinsic and intrinsic factors both physico-chemical and biological.

## Decomposition and Decomposers

### Process of Decomposition

Three stages of decomposition have been recognized;

 i) The formation of particulate detritus by physical and biological action;
 ii) The relatively rapid production of humus and release of soluble organics by physical and biological action;
 iii) The slower mineralization of humus.

Though decomposition results from both abiotic and biotic processes, the heterotrophic microorganisms are ultimately responsible for completion of the process. If decomposition does not takes place all the nutrients would remain tied up in dead bodies and no new life will be possible. Bacterial cells and fungal mycelia have sets of enzymes to carry out specific chemical reactions. These enzymes decompose organic matter and, some of the breakdown products are absorbed by the micro-organisms as

**Table 11.4: Biomass and net production of aquatic insect population from May 1978 to May 1979 (Roy 1986)**

| Date | Biomass* (g/haul) | Biomass (g/cubic meter) | Net Growth of the Population (g/cubic meter) | Period (days) | Net Production | |
|------|------|------|------|------|------|------|
| | | | | | (g/cubic meter/day) | (Kcal/ cubic meter/ day) |
| **1978** | | | | | | |
| 18 May | 0.92 | 1.862 | – | – | – | – |
| 19 Jun | 1.20 | 2.429 | 0.567 | 32 | 0.017 | 0.387 |
| 17 Jul | 1.50 | 3.036 | 0.607 | 27 | 0.022 | 0.565 |
| 16 Aug | 1.80 | 3.643 | 0.607 | 31 | 0.019 | 0.579 |
| 19 Sep | 2.00 | 4.048 | 0.405 | 35 | 0.012 | 0.590 |
| 16 Oct | 1.30 | 2.631 | -1.417 | 28 | -0.050 | -0.519 |
| 20 Nov | 0.80 | 1.619 | -1.012 | 35 | -0.028 | -0.266 |
| 18 Dec | 0.60 | 1.214 | -0.405 | 28 | -0.014 | -0.265 |
| **1979** | | | | | | |
| 15 Jan | 0.55 | 1.113 | -0.101 | 28 | -0.003 | -0.209 |
| 15 Feb | 0.71 | 1.437 | 0.324 | 31 | 0.010 | 0.227 |
| 20 Mar | 0.82 | 1.659 | 0.222 | 33 | 0.006 | 0.267 |
| 19 Apr | 0.95 | 1.923 | 0.264 | 29 | 0.009 | 0.355 |
| 17 May | 1.35 | 2.732 | 0.809 | 29 | 0.027 | 0.480 |

*1 haul = 494.0 litres of water.

food, and the secondary metabolites are released in the ecosystem. These secondary metabolites may be inhibitory or stimulatory in nature and have profound effects on the growth of other organisms in the ecosystem.

The rate of decomposition differs with different compounds, Fats, sugar and proteins are decomposed readily. But plant cellulose, lignin of wood and animal products like chitin, hair and bones take very long time to decompose. The more resistant products of decomposition end up as humus or humic substances which are the condensations of aromatic compounds (phenols) combined with the decomposition products of proteins and polysaccharides.

Further degradation of the complex phenolics by the microbes such as Bacteria (*Pseudomonas* and Acromobactor) and Fungi (*Aspergillus and Penicillium*) perform the degradation of the complex humic acid into simple phenolics. Ultimately, by the introduction of oxygen, ring fission may take place which finally release $CO_2$ following the path way as given by Tower and Subba Rao (1972) and Walker (1975). It has been depicted in the following modified way:

**Transaminase**

Phenylpyruval ←————————————————————————→ Cinnamic Acid

Phenylalamine

Ammonia-lyase

Microorganism

Phenylalamine

OH

Simple phenolics

Phenol

$CO_2$ ————→ Carbon

+$O_2$

Ring fission          Kreb's cycle          Humic acid

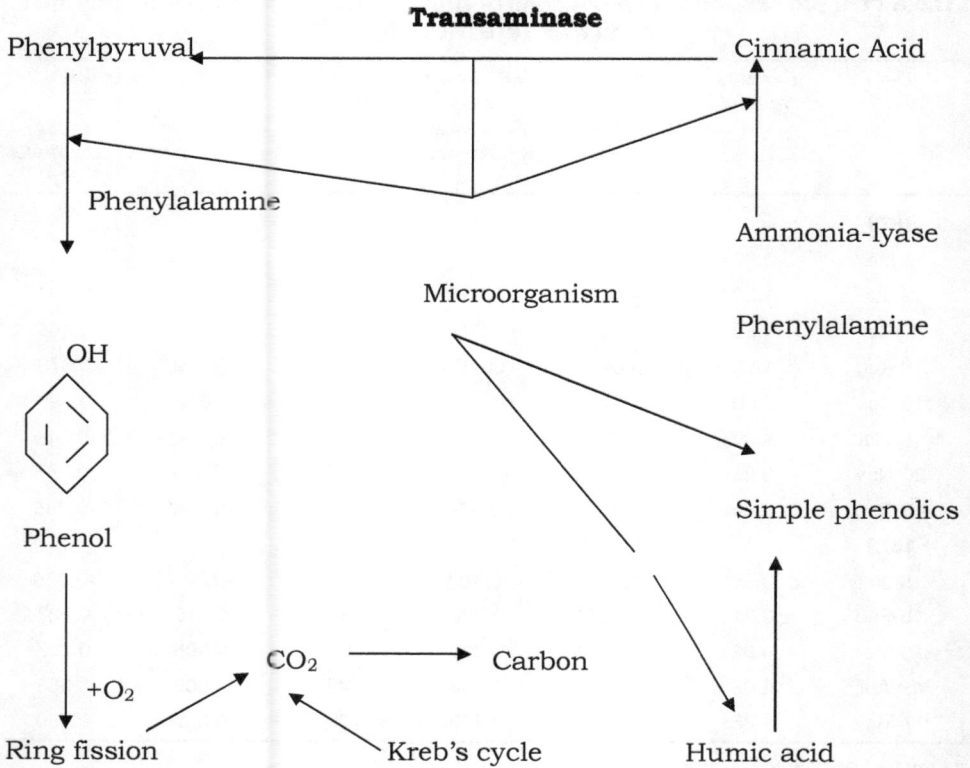

**Possible Degradation Pathway of Humus by Microorganisms**

In standing waters, like ponds and lakes most of the material incorporated in the organisms remain in the water and only a small fraction is deposited in the sediment. In flowing waters like rivers and springs part of materials are transporated out of the system. In aquatic systems bacteria are the most important agents in the process of organic decomposition, though fungi also act to a limited extent. Zooplankton also play an important part in the release of nutrients in water.

The decomposition of dead planktonic organisms proceeds very rapidly. Just after their death, zooplankton loose 21 per cent of its biomass, and after four hours the loss is 36 per cent and after one day 50 per cent. Rodina (1961) has shown that approximately 40 per cent of the residues of dead organisms are incorporated in the biomass of bacteria.

## Distribution of Bacteria in Freshwater

Bacteria in freshwaters can be determined quantitatively by:

i) Direct counting on membrane filters and employing scanning electron microscope;

ii) Indirect culture methods using solid or liquid media.

In direct method the inactive or dead bacteria and detritus particles are also included with active bacterial count. As a result the values of direct count gives thousand time greater values in comparison to indirect culture count. It is better to adopt culture method with nutrient concentrations that exist in natural environment.

The qualitative detection of certain bacteria is performed by means of specific colour reactions, metabolic tests and immunological procedures (Schmidt, 1973).

There is a relationship between the bacterial numbers and the trophic status of a water body which is shown in the Table 11.5.

**Table 11.5: Bacterial Numbers in the Surface Water of Lakes of Varying Trophic Status in Summer Time (Modified from Kuznetozov, 1970)**

| Lake Type | Bacterial Numbers (thousand per ml.) |
|---|---|
| Oligotrophic | 50–340 |
| Mesotrophic | 450–1400 |
| Eutrophic | 2200–12300 |
| Eutrophic reservoir | 1000–57900 |
| Dystrophic 430–2300 | |
| * Ganga river 1800–21100 | |

* Bilgrami and Datta Munshi (1986).

The generation time of bacteria varies from a few hours to several days depending on the species, availability of nutrients and other variable of water. Bacterial concentration in the epilimnion region shows close relationship with phytoplankton density, where they utilize the photosynthesis products and vitamins secreted from algae. Bacterial concentration is high in the metalimnion region and in the vicinity of water-sediment interface.

## Important Groups of Aquatic Bacteria

The metabolism of any organism whether micro or macroscopic requires its energy source, its electron donar and its carbon source for the manufacture of cell tissue.

Phototrophic bacteria use electromagnetic radiation as their energy source (light), while chemotrophic bacteria gain energy from oxidation reduction processes, with the transfer of electrons or 2H from the electrondonor to the electron acceptor. The initially reduced electron donor becomes oxidized and the electron acceptor is reduced. The capacity for utilizing inorganic electron only ($H_2$, $NH_3$, $H_2S$, $Fe^{++}$, $Co$, etc) is denoted as lithotrophy. When only organic electron donors are employed it is known as organotrophy. Autotrophic bacteria form their cell tissue originally by the fixation of $CO_2$. Heterotrophic bacteria manufacture their carbonates of cell tissue from organic compounds. So the bacteria are classified as chemo organotrophic, chemolithotrophic and phototrophic.

## Chemo-organotrophic (Heterotrophic) Bacteria

### Aerobic Carbon Assimilators

If any aquatic habitat lacks free oxygen, certain specific bacteria use combined oxygen from nitrites, nitrates and sulphates for the anaerobic oxidation of organic carbon compounds. Some bacteria are facultative anaerobes and can use both nitrate oxygen and also elemental oxygen, *e.g. Pseudomonas* spp., *P. aeruginosa, P. flucresens* and *P. denitrificans*:

Aerobic: $C_6H_{12}O_6 - 6O_2 \longrightarrow 6CO_2 + 6H_2O + \text{energy}$

Anaerobic: $5C_6H_{12}O_6 + 24 KNO_3 \longrightarrow 12N_2 + 24KHCO_3 + 6CO_2 + 18H_2O + \text{energy}$

The anaerobic route involves an organotrophic denitrification process, which is similar to removal of nitrate from water and effluents (Bringmann and Kuhu, 1967).

$C_6H_{12}O_6 + 3K_2SO_4 \longrightarrow 6KHCO_3 + 3H_2S + \text{energy}$

In addition to these reactions anaerobic decomposition of cellulose takes place by *Bacillus cellulosae dissolvens, Clostridium cellobioparus, pectin* by *Cytophaga, Flavobacterium, Pseudomonas* and *Bacillus* and *chitin* by *Bacterium chitinovorum*. All these anaerobic decomposition processes take place in the bottom sediments of natural waters.

### Chemolithotrophic Bacteria

Aerobic bacteria

Ammonium and Nitrate Oxidisers (nitrifiers)

*Nitrosomonas oxides ammonia to nitrate*:

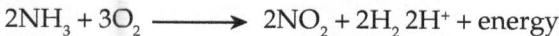

$$2NH_3 + 3O_2 \longrightarrow 2NO_2 + 2H_2\ 2H^+ + \text{energy}$$

*Nictobactor oxidizes nitrite further to nitrate:*

$$2NO_2 + O_2 \longrightarrow 2NO_2 + \text{energy}$$

Optimum conditions for nitrifiers in fresh water involve, a good supply of $O_2$, adequate buffering with calcium bicarbonate, neutral pH optimum temperature $8°C$ - $12°C$ and sufficient ammonia. In *oligotrophic waters* nitrifiers occur mainly at the sediment surface but in eutrophic lakes only in those zones which posses a good supply of oxygen. In flowing waters, especially when they are polluted with sewage effluent, nitrifiers are always present and thus heterotrophic nitrification as means of co-oxidation of ammonia with organic substances appears to be of much greater importance.

### Hydrogen Sulphide Oxidisers (Sulphurisers)

The microbial oxidation of $H_2S$ is carried out by colourless sulphur bacteria such as *Beggiatoa* and *Thiothrix* and *Thiobacteria (Thiobacillus)*:

$$2H_2S + O_2 \longrightarrow 2H_2O + 2S + \text{energy}$$

The energy obtained from the reaction is used for the assimilation of $CO_2$:

$$6CO_2 + 12H_2O \longrightarrow C_6H_{12}O_6 + 6H_2O + 6O_2$$

## Anaerobic Chemolithotrophic Bacteria

### Nitrare-Reducers (Denitrifiers)

Nitrate reduction invariably plays a part in thise water bodies where nitrate is present within the redox-potential range of biological reduction processes. For denitrification nitrate plays the part of the electron acceptor; electron donors are either elemental hydrogen as for the facultative chemo lithoheterotroph *Paracoccus dentrificans*:

$$5H_2 + 2KONO_3 \longrightarrow N_2 + 2KOH + 4H_2O + energy$$

or

Oxidisable inorganic sulphur compounds which are oxidized by obligate chemoautotrophic *Thiobacillus denitrificans*:

$$5S + 6KNO_3 + 2H_2O \longrightarrow 2N_2 + 3K_2SO_4 + 2H_2SO_4 + energy$$

### Sulphate-Reducers (Desulphurisers)

Some strains of Desulfovibrio desulfuricans are facultative Chemolithotrophs and reduce sulphate, sulphite, bisulphate, thiosulphate and tetrathionate with elemental hydrogen and a gain of energy:

$$4H_2 + K_2SO_4 \longrightarrow K_2S + 4H_2O + energy$$

### Methane Formers

Methane-forming bacteria convert activated hydrogen and carbon dioxide using hydrogenase.

$$4H_2 + CO_2 \longrightarrow CH_4 + 2H_2O + energy$$

In this wat *Methanobacterium Methanosarcina Methanobacillius* and *Methanococcus* spp. are able to obtain energy.

## Phototrophic Bacteria

Phototrophic bacteria include *Chromatiaceae (purple sulphur bacteria)*, *Rhodospirillaceae (sulphur-free purple bacteria, photo-organo heterotrophs)*, *Chlorobiaceae (green sulphur bacteria)* and *chloroflexaceae (chloroflexus* group). The chromatiaceae consists of the taxa *Thiocapsa. Thiospirillum, Chromatium* and *Lamprocystis; Chlorobiaceae* of *Chlorochromatium, chlorobium, Pedodictyon* and *Chlorobacterium; Rhodospinrillaceae* of *Rhodoseudomonas (rods), Rhodospirillum (spirilla)* and *Rhodomicrobium* and *Chloroflexaceae* of *Chloroflexus. Rhodospirillaceae* contain assimilatory pigments bacterial *chlorophyll a (+b), Chlorobiaceae, bacterial chlorophylls a,b,d* and *e*, and *Chromatiaceae the chlorophylls a and b* (*Good win, 1966* and *Schlegal, 1981*).

The Chromatiaceae use $H_2S$, which is oxidized to elemental sulphur or to sulphuric acid:

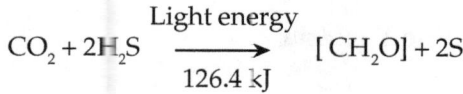

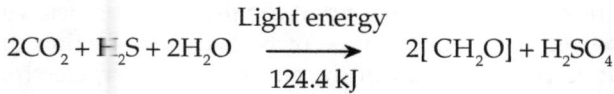

$$CO_2 + 2H_2S \xrightarrow[126.4 \text{ kJ}]{\text{Light energy}} [CH_2O] + 2S$$

$$2CO_2 + H_2S + 2H_2O \xrightarrow[124.4 \text{ kJ}]{\text{Light energy}} 2[CH_2O] + H_2SO_4$$

The Chlorobiaceae oxidize chiefly $H_2S$ but also elemental sulphur to sulphuric acid:

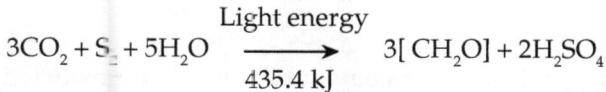

$$3CO_2 + S + 5H_2O \xrightarrow[435.4 \text{ kJ}]{\text{Light energy}} 3[CH_2O] + 2H_2SO_4$$

Bacterial photosynthesis proceeds anaerobically and does not generate oxygen (anoxygenic photosynthesis). In natural waters therefore photo-autotrophic sulphur bacteria frequently occur at the boundary layer of oxygen and hydrogen sulphide, where sufficient light energy still penetrates (Gorlenko and Kusnezov, 1972).

The Rhodospirillaceae are also anaerobic, but depend on organic electron donors;

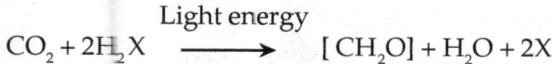

$$CO_2 + 2H_2X \xrightarrow{\text{Light energy}} [CH_2O] + H_2O + 2X$$

*where,* x = organic matter.

Rhodospirillaceae occur on oxygen-free zones and with help of special assimilatory pigments are capable of using light of very long wave lengths for photosynthesis. At the same time they are also active at greater depths where they can utilize the blue or greenish-blue end of the spectrum.

## Aquatic Fungi

It has already been mentioned that besides bacteria, to a limited extent fungi also take part in organic decomposition in aquatic system (*Hohnk, 1962*). But mostly these fungi are not the true inhabitants of water but is added to the system from outside with decaying woods, twigs and leaf litters. Only 15-20 per cent of the total fungal population is aquatic. This is because of fungi unlike bacteria can not tolerate much depletion of oxygen content which sometimes may occur in aquatic systems. Members of *Chytridiales* and *Saprolegniales* play important role in cellulose decomposition in freshwater. Freshwater fungi constitute two main groups.

  i) Zoosporic water moulds of which *Saprolegniales* are the dominant, and
  ii) *Conidial Hyphomycetes*; some of which can flourish in moderately polluted freshwaters.

Fungi have been reported from the river Ganga (*Bilgrami and Datta Munshi, 1985)*:

CHYTRIDIOMYCETES          Nil

OOMYCETES

   *Saprolegniales.*

      *Achlyaprolifera* (Nees) de Bary

      *Isoachlya anisospora* var. indica Saks, set Bharg

      *Saprolegniax ferax* (Gruithuysen) Thruet

   *Peronosporales*

      *Pythium butleri* Subram

      *P. carolinianum* Mathews

ZYGOMYCETES

   *Mucorales*

      *Absidia spinosa* Lendner

      *Actinomucor* sp.

      *Circinella simplex* Van Tiegh

      *Mucor mucedo* (Linne) Brefeld

      *Syncephalastrum recemosum* Cohn ex Schoroter

PYRENOMYCETES

   *Sphaeriales*

      *Chaetomium funicola* Cooke

      *C. perpulchrum* Ames

      *Diaporthe* sp.

      *Podospora mannopodalis* Cains

DISCOMYCETES

   *Pezizales*

      *Ascobolus* sp.

LOCULOACOMYCETES

   *Pleosporales*

      *Leptosphaeria aquatica* (Tilak and Kulkarni)

      *Leptospora rubella* (Pers) Rabenh

      *Ophiobolus herpotrichus* (Fr.) Sacc.

      *O. spirosporus* Ahmad

      *Thyridium* sp.

COELOMYCETES

  *Sphaeropsidales*

      *Phoma nebulosa* (pers ex. Fr.) Berk

HYPHOMYCETES

  *Moniliales*

      *Alternaria gomphrenae* Togashi

      *Aspergillus fumigatus* Fr.

      *B. japonicus* K. Saito

      *A. phoenics* (Corda) Thom

      *Drechslera australiensis* (Bugnicout) Subram and Jain

      *Nigrospora oryzae* (Berk and Br.) Petch.

      *Penicillium citrinum* Thom

      *P. herquei* Bain, and Sart

      *P. pinophilum* Hedgcock

      *P. spiculisporum* Letman

They occur on plant parts and animal tissue. Most common fresh-water fungi are *Saprolegnia, Achlya, Aphanomyces, Dictychus, Leptolegnia, Allomyces* and *Rhizophydium. Cryptococcus* is the most common aquatic yeast. Fungal parasites on fishes are quite common.

## Recycle Pathways and Recycling of Nutrients in Freshwater

It has been shown that though bacteria and fungi are the major agents in nutrients regeneration, there are some other important ways of recycling (Figure 7.2). In recycle pathway microorganisms as well as small animal detritivores are responsible for nutrient regeneration in soils. In aquatic habitats where heavy grazing of phytoplankton by Zooplankton and fishes take place, the recycling mostly occurs through animal excretion.

Zooplankton release directly the dissolved organic compounds of phosphorus, nitrogen, $CO_2$, and other vital nutrients, which can be utilized by the producers. Other pathways are involved in the mobilization of resources from abiotic reservoir to biotic cycle.

Another important way of nutrient regeneration in aquatic environment is the autolysis by cellular enzymes (self-dissolving) of dead plants, animals and their fecal pellets without the action of microorganisms. This does not involve any metabolic energy. In aquatic system where dead bodies are small exposing a large surface/volume ratio autolysis process may contribute 15 to 75 per cent of the nutrient release, before the microbial attack begins.

## Energetics

Energy is the ability to do work. The sun is the source of all energy for living organisms in this earth. In sun continuous nuclear rearrangement by transformation of hydrogen atoms into helium releases fantastically high amounts of energy. This energy radiates out in all directions in the form of electromagnetic waves called solar radiation.

## Nature of Energy and Laws of Thermodynamics

There are different forms of energy such as radiant energy, light energy, heat energy, photochemical energy and mechanical energy. The wavelengths between 0.4 to 0.7 µ of radiant energy is the visible light energy. Heat energy is the other form of radiant energy which warms the earth, heats the atmosphere and drives the water cycle. Photochemical energy is fixed in carbohydrates and other compounds by photosynthetic activity of green plants. Mechanical energy exists in two forms – the potential and kinetic.

The behaviour of energy is governed by the following laws:

i) First law of thermodynamics or the energy conservation law states that energy may be transformed from one type into another but is neither created nor destroyed. For example light energy can be transformed into heat or potential energy of food but none of it is destroyed.

ii) Second law of thermodynamics or the entropy law may be stated as follows:-

No process involving energy transformation will spontaneously occur unless there is a degradation of the energy from a concentrated form into a dispersed form. For example heat energy from a hot object will spontaneously tend to become dispersed into the cooler surroundings. The second law of thermodynamics also states that: As some energy is always dispersed into unavailable heat, no spontaneous transformation is 100 per cent efficient. As for example when light energy is transformed into potential energy of photoplasm only a portion of it is transformed and rest of it remain unavailable. Entropy is a measure of unavailable energy resulting from transformation. This term is also used as a general index of the disorder associated with energy degradation. Ecosystems are open thermodynamic systems which can maintain a steady state as long as there is a constant input of matter and free energy and a constant out flow of entropy in the form of heat and waste.

## Units of Energy

Energy is measured in several units. The gram calorie (gcal) is the amount of heat necessary to raise temperature of 1 gram of water to 1°C. It is the most convenient unit of energy. When large quantities of energy are involved the kilogram calorie (Kcal) is used. When energy flow for a given area is concerned it is measured per square centimeter or square meter. Recently joule unit is used which is equal to 0.24 gram calorie.

## Ecological Energetics

It concerns mainly with the quantity of energy reaching an ecosystem per unit area per unit time, the quantity of energy trapped by green plants and its conversion to organic matter. An ecologist is interested in finding out the quantity of energy and the path of energy flow through the trophic levels over a period of time in a known area.

According to Philipson (1966) 15 x $10^8$ calories/metre$^2$/year of solar energy enters earth atmosphere. About 33 per cent of the incoming solar radiation is reflected by clouds and another 9 per cent by suspended dust particles. Ozone layer, water vapour and other atmospheric gases held another 9 per cent of the radiation. Ultimately 47 per cent of it reach the surface of the earth. Quantity of energy received is further dependent upon latitude of the place. The equatorial region receive maximum energy. From the works of several scientists this has been found to be 2.5 x $10^8$ cal/ $m^2$/yr in Britain (Phillipson 1966), 4.7 x $10^8$ cal/$m^2$/yr in Michigan and 6.0 x $10^8$ cal/ $m^2$/yr in Geogia, U.S.A. (Golley 1960) and 7.5 x $10^8$ cal/$m^2$/yr at Varanasi, India (Singh 1967). About 1 to 5 percent of this energy is converted to chemical energy of organic matter by the green plants and rest is lost as heat.

The energy stored in the organic matter of green plants is passed through the ecosystem in a series of steps of eating and being eaten known as the food chain. Within any ecosystem there are two major food chains, the grazing food chain and the detritus food chain. In deep-water aquatic ecosystem with their low biomass, rapid turnover of organisms and high rate of harvest, the grazing food chain is dominant.

In most terrestrial and shallow water ecosystems (Swamps and marshes), with their high standing crop (see primary productivity) and relatively low harvest of primary production, the detritus focd chain predominates. According to the concept of ecological energeties the *energy always flows in one direction along the food chain* in any ecosystem. Energy which is accumulated by the green plants is primary production. After consumption by the consumers, it is either diverted to their maintainnnnance, growth and reproduction or is passed from the body as faces and urine.

## Trophic Levels and Ecological Pyramids

Each step in the food chain represents a trophic level. The food energy is transferred from producers through series of consumer organisms in a food chain. The first trophic level belongs to the producers, the second to the herbivores, the third to the first-order carnivores and so on. Animals at the lower level, such as the Zooplankton may occupy a single trophic level, but most of the animals at the higher levels like fishes participate simultaneously in several trophic levels forming a complex food web. Thus in nature food chairs can not operate as isolated sequences, but are interconnected with each other forming some sort of interlocking pattern which is referred to as a food web (Munshi, Singh and Singh, 1990; Munshi *et al.*, 1991) (Figure 11. 1 a-n).

Trophic levels as considered by most ecological references do not include the decomposer and parasites, but these should be considered as herbivores or carnivores depending upon the nature of their food source. Decomposers feeding on the dead plant material, as well as bacteria occupying the rumen of ungulate animals (Cows and buffaloes), should be considered functional herbivores. Decomposers feeding on the dead bodies of animals should be considered carnivores. Likewise all the various steps in energy transfer in an ecosystem can be placed in some trophic level.

By adding all of the biomass or living tissue contained in each trophic level and all of the energy transferred between levels, one can construct pyramids of biomass and energy for the ecosystem.

## Ecological Pyramids

The pyramid of accumulation of organic matter (biomass) is large, where life cycles are long and the rate of harvesting is low. In an aquatic ecosystem the pyramid of biomass is inverted. The primary production is concentrated in microscopic algae (phytoplankton). These algae with a short span of life, multiply rapidly, accumulate little organic matter and are harvested heavily by the zooplankton. As such at any point, the standing crop is low giving rise to smaller base of the pyramid.

When pyramid of energy is constructed it indicates not only the amount of flow at each level, but the actual role the various organisms play in the transfer of energy. The base upon which the pyramid of energy is constructed is the rate at which food energy passes through the food chain (Figure 7.3). Some organism may have a small biomass, but the total energy they assimilate and pass on may be greater than that of organisms with a much larger biomass. In the pyramid of biomass, these smaller organisms would appear much less important in the community than they are. Thus pyramid of energy, gives the true picture of the functional role of the organisms in an ecosystem.

These graphic representations of production and transfer always form pyramids in accordance with the second law of thermodynamics. Less energy is transferred from each level to the next one that the input into it. Especially in open-water communities where the producers have less bulk than the consumers, the energy they store and pass on must be greater than that of the next level. Otherwise the biomass that producers support could not be greater than that of the producers themselves. This energy flow is maintained by a rapid turnover of individual plankton, rather than by an increase of total biomass.

Another pyramid commonly found in ecological literature is the pyramid of numbers (Figure 7.3), which was advanced by Charles Elton (1972). He described that there are much differences in the numbers of organisms involved in each step of the food chain. The animals at the lower end of the chain are the most abundant. The number of carnivores decrease in successive links to a few at the top. The pyramid of numbers ignores the biomass of organisms or energy transferred by them.

## Model of Energy Flora

The energy flow model was first proposed by Lindeman (1942) which was based on the assumption that the laws of thermodynamics are also applicable to plants and animals. According to him plants and animals can be grouped to different trophic levels according to their feeding habits. There are at least three trophic levels in any ecosystem *viz.*, producers, herbivores and carnivores and the equilibrium exists in ecosystem.

In the Linderman trophic-dynamic models the energy content of standing crop of any trophic level is designated by the Greek Capital Lambda. A. This letter is followed by a numerical subscript to denote trophic level. Thus $A_1$ represents producers; $A_2$ that of herbivores and so on. $A_n$ indicates any designated trophic level. Energy is continuously entering and leaving (the dynamic aspect), from one trophic level to another in course of time. The energy flow is indicated by a lower-case (small) Lambda, â. Thus $â_1$ represents the proportion of energy received by the organisms of any one trophic level $\Delta_n$ receive from the trophic level below. For any trophic level designated as $\Delta_n$, $â_n$, represents $â_{n+}$ + R and $â_{n+1}$ is the loss of heat due to respiration.

Following Linderman's nomenclature, Phillipson (1966) has given a generalized equation for expressing the rate of change of energy content of any trophic ($\Delta_n$) level:

From this equation it is clear that the rate of change of the energy content of a trophic level is equal to the rate at which energy is assimilated minus the rate at which energy is lost from it. In this equation $â_b$ id positive and represents the contribution of energy from the previous trophic level, $\Delta_{n-1}$ and $â_n$, represents the sum of energy lost from $\Delta_n$.

Lindeman worked in Ceder Bog lake of Minnesota, where he estimated that out of total 118,872 cal/cm$^2$/yr incident solar radiations, the primary producers could capture only 111 cal/cm$^2$/yr which is less than 0.1 per cent. Energy passed on the herbivores was only 15 calories and carnivores only 3 calories.

This energy capturing efficiency value is very low is comparison to other works in different ecosystems in recent years.

Accurate measurement of energy transfer efficiency was first determined by Odum (1957) in Silver Springs and other springs of Florida (Figure 7.6). Later on Odum (1963) was able to develop a basic model of energy flow for three trophic levels (Figure 7.7).

$$\text{Energy fixation efficiency} = \frac{\text{Net annual production [kcal m}^{-1}\text{yr}^{-1}]}{\text{½ Solar Radiation [kcal m}^{-1}\text{ yr}^{-1}]} \times 100$$

Or

$$\text{Energy transfer efficiency} = \frac{\text{Net annual production [kcal m}^{-1}\text{yr}^{-1}]}{\text{Energy input [kcal m}^{-1}\text{ yr}^{-1}]} \times 100$$

In this lotic system we see that out of the annual total amount of incident energy (L) of 7.1 x 10$^6$ kJ/m$^2$ an amount of 17.2 x 10$^5$ KJ (24 per cent) was absorbed ($L_A$) by the producers. The producers in this system consists almost of higher aquatic plants, the gross productivity ($P_G$) of which amounts to 8.7 x 10$^4$ kJ. The trophic energy intake (or Lindeman's) efficiency or transfer efficiency ($P_G/L$) is 1.2 per cent of the assimilated amount of 8.7 x 10$^4$kJ, 57.5 per cent was respired and only 3.4 x 10$^4$kJ were accounted for by net productivity to form the potential energy supply for the consumers. The conversion efficiency at the producers level (net productivity/assimilation or $P_1/A_1$) is only 0.42 of this 3.4 x 10$^4$ kJ, the largest amount was transported down stream and only 1.4 x 10$^4$ kJ was actually assimilated by the primary consumers, of which an amount of 2.0 x 10$^3$ kJ/m$^2$ was imported from outside. The energy flow from the producer level to the primary consumer level ($l_t/l_t$) was thus only 16 per cent. Out of 1.4 x 10$^4$kJ at the primary consumer level, only 1.6 x 10$^3$ kJ (11 per cent) was transferred to the secondary consumers and this only 87.9 kJ (6 per cent) reached ther terminal consumers (Crocodiles). From total of gross productivity 88 per cent was used up in the ecosystem itself and 12 per cent was carried downstream as detritus and so lost. Depending upon energy input from the outside, lotic ecosystems differ substantially from lakes and from terrestrial ecosystems whose main source of energy is primary production within the system. Energy is supplied mainly by allochthonous sources in lotic systems. Energy may come in form of leaf and branch litter fallen from the vegetation growing on the catchment areas which is carried to the system by wind or rain water. Like this nutrients may come to the system from the dumping of urban and industrial effluents into the flowing water.

There is approximately 50 per cent absorption of which 1 per cent conversion of light takes place at the first trophic level. Secondary productivity ($P_2$ and $P_1$) is about 10 per cent at successive consumer trophic levels, although the efficiency tends to be higher, say 20 per cent at the carnivore levels as shown in the diagram. This explains the reason for limited number of links in a food chain.

## Comparative Energy Flow in Aquatic and Terrestrial Systems

By constructing 2-channel energy flow model (Odum 1963) for two contrasting types *i.e.* aquatic and terrestrial systems it has been found that in aquatic community the energy flow via grazing food chain is larger than via detritus chain. But in forest ecosystem 90 per cent or more of the energy flows through detritus food chain. But if we go for further details then we find that the amount of energy flow through food chains does not always strictly follow this rule. In a heavily grazed pasture or grassland 50 per cent or more of the energy flows through the grazing food chain while in aquatic system like marshes and wetlands most of it passes through detritus food chain. Even in aquatic system like ocean, through large amount of energy flows via grazing food chain but after trophic level II, it is diverted to the detritus food chain in form of unassimilated organic matter as zooplankton graze more phytoplankton than they can assimilate.

# Epilogue–
# Culture of Commercially
# Important Fishes

## Problems and Cultural Procedures

The commercially important air-breathing fishes of India are *Heteropneustes* (singhi), *Anabas* (Kawai), *Clarias* (magur) and *Channa* (murrel). They are held in some esteem as food fishes and are generally found in the swamps of Bihar, Bengal, Assam and some parts of South India. These fishes are endowed with remarkable powers of respiration that enable them to lead an amphibious life and to survive the adverse condition in swampy areas. The extensive swamps, subject to the ravages of the seasons, permit the survival of only a limited number of species. Reclamation of the swamps for carp culture would entail considerable expenditure, yet what a waste of valuable food sources if these vast areas are not better utilized for the culture if air-breathing fishes. The importance of the problem of culturing and propagating these fishes in swamps needs to be recognized and, subsequently, techniques developed for the proper utilization of swamps for these purpose. Realisation of these objectives would have considerable significance in the augmentation of fish production especially in the rural areas (Dehadrai, 1972, 1978; Dehadrai and Thakur, 1980).

*Anabas, Clarias, Heteropneustes* and *Channa* may be cultured separately or together. A carp culture nursery pond could be used for culturing these fishes from November to May. Besides this, discarded water bodies of shallow water could also be used. In North Bihar these fishes are cultured in derelict ponds where "makhana" (Euryale) and "singhara" (Trapa) are also cultured. Culturing these fishes in small tanks at home is also feasible.

## 1. Cage Culture

In reservoirs or any other large water bodies, the above-listed air-breathing fishes can be cultured in split bamboo mat cages of 152cm x 76cm size. The half portion at the top is kept open while the lower half is covered with a net cloth. The cage is usually supported by bamboo poles and kept half immersed in the water so that its bottom does not touch the bed of the water body. About 100 to 200 juveniles/square metre should be stocked in a cage.

For monoculture, juveniles ranging from 5 to 10 cm should be stocked at a density of 1 to 5 lakhs/ha. About 25 to 30 thousands/h may be cultured along with carp in ponds. However only 50-100 juveniles/square metre may be stocked in tanks.

Proper supplementary feeding is required for successful culture and high production of artificial feed comprising a mixture of rice bran and fish meal, dried prawn or silkworm pupae, cow dung, kitchen refuse and slaughterhouse waste at a rate of 3 to 5 per cent of the stocked fished weight may be used as supplementary feed for the juveniles.

Collections of seeds are always a problem in the culture of the fishes. However, the seeds of *Anabas, Clarias* and *Heteropneustes* may be obtained by two methods.

    i) By the hypophysation technique,*i.e.*by pituitary extract injection and

    ii) From nature during the monsoon period (June-September).

## 2. Production of Eggs and Fry (Seed)

Experiments were carried out by the hypophysation technique to produce seed (=fertilized eggs or spawn) and fry of *Anabas, Clarias, Heteropneustes* and *Channa* on a mass scale. The fishes were injected with varying doses of pituitary in acquaria, plastic pools breeding pits, paddy fields and shallow ponds.

# Induced Breeding in *Heteropneustes fossilis*

Ramaswamy, L.S. and Sunderaraj, B.I (1957) were the first to breed *Heteropneustes* through hypophysation. The dose was 1.5 homoplastic pituitaries from male donors and 0.5 pituitary from female donors. The fish also responded to heteroplastic pituitary. The pituitary glands of major Indian carps (Catla, Labeo, Cirrhinus) proved effective and economical, the routine dose being 8-10 mg of gland/100 weight of the recipient. The glands were used to prepare an aqueous extract for injection.

# Induced Breeding in *Clarias bactrachus*

*C. batrachus* does not attain proper ripeness in captivity it is reared on a suitable supplementary diet. Stocked at the rate of 40 fish/m² and fed on a diet-bran and crushed aquatic insects or fish meal (ground dried fish) in the portion of 1:1 works fairly well for the upkeep of spawners. However, Spawners reared in a natural pond give better results than those reared in plastic pools or cages.

For induction of spawning in *Clarias*, the pituitary glands of major Indian carps, with doses varying from 5-20 mg/100 g body weight of the recipient, are generally used. Intramuscular injections are given slightly above the lateral line in the middle region of the body. Keeping in view the monogamous habit of the fish, pairing is made in the ratio of 1 male, 1 female. Spawning activity starts 16-20 hr after the injection and lasts for 6-12 hr. A female of 150-200gm produces over 4,000-5,000 eggs. The best results are obtained when the spawners after pituitary injection are released in small plots of paddy fields. Unlike *Heteropneustes* it is difficult to induce breeding in *Clarias* under laboratory conditions. Even a flow water system installed in a river failed to induce *Clarias* to breed.

## Natural Breeding

Successful mass breeding of *Clarias* was achieved in a tank of 0.4 ha having embankments on all sides with a breeding tank of 0.1 ha situated inside it. The tank received a natural flow of water from an upand catchment area. Hundreds of horizontal burrows 30 cn deep and 8 cm wide were made on the lower portion of the inner walls of the embankments on all sides and aquatic weeds were provided inside the burrows. The water level was maintained at 1.5-2 in above the mouth of the burrows. A total of 100 each of males and females of the fish, with an average llength/ weight of 27.5cm/150gm were released. The fingerling production was estimated to be about 1.1 lakh.

In several other countries, such as the Philippines and Thailand, *Clarias* culture is very prevalent. Micha (1972) has reported successful induced breeding at the University of the Philippines in large aquaria. Of the total number of breeders injected. 50 per cent responded well; the number of fertilized eggs varied between 2000-7000.

In Thailand, *C. batrachus* is breed in specially prepared spawning ponds under simulated ecological conditions. Regular spawning ponds 1-3 ha. Size are constructed close to irrigation canals. Along the margin of the pond, a channel 2-3 on wide and 1-2.5 m deep in dug out for the fish. In the middle portion of the pond, several rows of about 2m wide raised platforms are constructed by digging several small channels of 50 cm wide in between them. Grasses are allowed to grow on these platforms and 100-120 burrows of 20-25 cm in diameter and 15-20 cm deep are dug to provide nesting space for the brood fish to lay eggs.

With the advent of the spawning season, the breeding operation is started. Feeding is stopped and water is pumped in to inundate the burrows contracted on the raised platforms. The level of the water is raised 25-35 cm above the burrow's holes. The fish breed in the channel are stimulated by this inundation., move out to the spawning area and breed within 2-3 days. After hatching, the larvae stay in the burrows, which are harvested 9-10 days later. Each next yields 2000-5000 fry.

## Nursing of Fry and Fingerlings

Nursery ponds are usually 800-3,500 m$^2$ water area, with a depth of 50 cm. in the middle of the pond a channel 2 m wide and 50-60 cm deep is constructed to facilitate collection of fry during investing. The nursery pond is fertilized with manure at 30kg/100m$^2$ to produce a bloom of zooplankton, stocking is done at 250-350 fry/m$^2$. Artificial supplementary feeding is started from the second day of stocking with ground trash fish and rice bran mixed in the ratio of 9:1 and fed at 1 kg feed/1,000 fry every 2-3 days. If the fry are to be collected at a very young stage, the initial feeding is done with cooked chicken eggs inside the hapas, before the fish are released into the nursery pond.

Aerial respiration in *C. batrachus* starts on the 10th -11th day of development, when the post-larva begins vertical trips by wriggling swiftly through the water column to reach the surface. For the return trip the larva just passively sinks; however, its body is vertically disposed with the head up and the laid down. This vertical disposition is actually caused by air-filled accessory organs, which impart buoyancy (Quasim, Quayum and Garg, 1960). Post-larvae feed actively on zooplankton but also accept aryificial food.

## Culture of *Channa*

All the three species of *Channa, C.marulius, C. striatus* and *C.punctata* have been induced to breed for production of fry for controlled culture. Doses of 15-40 mg/kg body weight of carp pituitary for males and 80-120mg/kg body weight for females are recommended. The hatchings emerge after 24 hours of injection and require intensive care. Rearing can be done in the hatchery with an artificial feed of boiled chicken egg yolk. One egg yolk is sufficient for 50,000 hatchings/day. After a week, the larvae are ready to take zooplankton, ponds with a low water column are suitable for caring fingerling of *C. marulius, C. punctata, C. gachua* and *H. fossilis* in the proportion of 30:1.5:1:2 at 34,500 fingerling/ha (Devraj, 1973, Jhingran, 1984)

Harvesting of murrels poses problems since the fish burrow down into the bottom muddy and avoid nets and other gear. Cage Culture of *murrels* is possible with supplementary feed. An adequate population of forage fish such as *Puntius sophore* and *Oxygaster bacila* is maintained in the tank, keeping in view a proper predatory-prey ration.

## Enemies of Fry and Management Measures

The larvae of air-breathing fishes are generally attached by copepods, especially *Cyclops sppi*. And therefore the zooplankton feed should be sieved before it is released into the rearing tanks.

*Notoncctids* (708 mm long) were found to attack and kill fry of *C. striata.* Dragonfly and May fly larvae were found to prey upon the fry and fingerlings of murrels. To overcome this problems, the same precautionary measures were adopted for controlling aquatic insects. These included time treatment of the soil during preparation of the nursery ponds and spraying of insecticides until the murrel fry reached a size 22-24 mm.

## Cannibalism and Management Measures

Cannibalism was found to be very pronounced in *Channa, Heteropneustes* and especially in the fry and fingerling stages; the larger and stronger prey upon the weaker and smaller. For this reason, the fry-rearing phase in these fishes, unlike in carps, is complex. Fry and fingerlings are separated according to size by means of a sieve, transferred to different plastic pools and provided feed in abundance.

In the field, on a commercial scale, the hatchings of *Heteropneustes, Clarias, Anabas* and *Channa* are reared separately after yolk absorption in hapas of nylon cloth with

50 mesh/linear cm fixed in ponds or swamps. Separate rearing ensures abundant zooplankton for the aerial breathing of baby fishes. The management can also manipulate fish population density. It is now possible to achieve a 50-75 per cent survival of post larvae of *Heteropneustes* to a size of about 22mm, which is the best size for stocking in swamps in cages (Dehadrai, 1972).

## Production of Air-breathing Fishes in India

Production of 320kg/ha/3 months was reported after stocking the induced breed stockable material of Heteropneustes in a swampy water in Bihar. While *Heteropneustes* is compartible with *Clarius* and *Anabas*. It has considerable potentiality for monoculture under semi-intensive systems (Jhingran, 1984).

Intensive culture of *C. batrachus* in recirculating – filtering ponds has been initiated at the Central Inland Fisheries Research Institute, Barrackpore, West Bengal. At a stocking density corresponding 1.1 million fingerlings/ha in numbers and 15 t/ha in weight, a standing crop of 39 t/ha has been computed in a 90 days culture period. The rate of survival is 96 per cent. Fish are fed on semi-dried balls consisting of 70 per cent fish meat, 25 per cent groundnut oil cake and 10 per cent wheat flour as binder, fortified with essential vitamins and mineral mixtures (Jhingran, 1984).

At the Bhadra centre (Karnataka, South India) an All India Co-ordinated Project on Air-breathing fishes under CIFRI, as much as 895 kg/ha/yr production of *C. Straita* was achieved from Swampy waters by feeding it with a self-generating stock of minnows and other forage fishes.

At Dharbhanga (North Bihar) centre of the Project, a mixed culture of *Clarias*, *Heteropneustes* and *Anabas* stocked at the rate of 25,000/ha in a dereliet pond (0.4 ha), gave a production equivalent to 1,200 kg/ha in 7 months without fertilizer and supplementary feed (Dehadrai, 1972a). Although *Clarias* is known to be carnivore in nature, nevertheless when stocked in slit-laden swampy ponds it subsists on bottom detritus.

Another experiment of integrated aquaculture of *Clarias, Heteropneustes* and *Anabas* with a total stocking rate of 70,000/ha fish along with Makhana (*Euryale ferox*) yielded a fish production of 1,200 kg/ha in 8 months and 320 kg of Makhana seeds (used as food) which is a highly prized aquatic cash crop. Paddy-cum fish culture has been practiced in India since time immemorial (Dehadrai and Mukhopadhya, 1979). In areas where paddy fields retain water for 3 to 8 months in a year paddy-cum-fish culture can provide an additional supply of fish crop. The culture of fish in fields, which remain flooded even after the paddy is harvested, might also serve as an off-season occupation of farmers. In India carps along with common *Murrel, Channa straiatus* and minnows are used for the purpose.

The utilization of paddy fields for fish culture actually comes in direct conflict with the use of land resources and the future prospects of paddy-cum-fish culture are not bright unless the agriculturists make terms with pisciculturists and use only those pesticides which will effectively combating infestations are also tolerant to fish.

The traditional carp nurseries and rearing ponds, after harvesting these fingerlings by November/December, could be used for the culture of *Clarias*, *Heteropneustes* and *Anabas* from January to June every year, and their marketable stock harvested before the monsoon to permit carp seed production thereafter (Dehadrai, 1978).

The air-breathing fishes discussed in this book are known for their nutritive, invigorating and therapeutic qualities and are recommended by physicians for inclusion in the diet of convalescents.

# Reference

P.V Dehadrai (1972,1978)

P.V. Dehadrai and Thakur, 1980

Qasim S.Z.

Ramaswamy and Sundaraj 1957

Micha, T.C. 1972

Qasim S.Z. 1960

Jhingran, V.G., Natarajan, A.V., Banerjee S.M, and David, A. (1969)

# References

Agrawal, H.P. (1980): Some observation on the seasonal variation in the gonads of *Indonale caerules* (Les) (Molluscs: Unionidae) – Bull, Zool. Surv. India, **3(1-2):** 87-92.

Ahmad, M.R. (1967): Algal flora of some ponds of Kanpur. Hydrobiol. **29(1-2)** 156-164.

Alikunhi, K.H. (1057): Fish culture in India. ICAR Farm Bull. 20.

Atkins, W.R.G.and G.T. Harris (1924): Seasonal changes in the water and *heleoplannkton* of freshwater ponds, Sc. Proc. Roy, Dub. Soc. (N.S.) **18,** 1-21.

Bauer, G. (1987): Reproductive strategy of the freshwater Pearl mussel *Margaritifera margaritifera – Biol. Conserv., **37:** 691-704.*

Bloomer, H.H. (1930): A note on the sex of *Anodonta cygnea* – Proc. Malac, Soc. Lond., **18(1)**: 10-14.

Bloomer, H.H. (1931): *A note on the anatomy of Lamelidens marginalis Lamarck and L.thwaitesii Lea – Proc.malac, Soc. Lond., **19(1)**: 10-14.*

Bloomer, H.H. (1934): *On the sex and sex modification of the gill of Anodonta cygnea – Ibid,* **21(1)***: 21-28.*

Bloomer, H.H. (1935): *A further note on the sex of Anodonta cygnea-Ibid,* **21(5)***: 304-321.*

Bloomer, H.H (1939): A note on the sex of Anodonta – Ibid, **23(5)**: 285-297.

Brock, T.D. (1967a): Life at high temperature. Nature **214**:882-885.

Brock, T.D. (1967b): Relationship between standing crop and primary productivity along a hot spring thermal gradient. Ecology **48**; 566-571.

Brock,T.D. (1967c): Micro-organisms adapted to high temperatures. Nature **214**: 882-885.

Brooks, J.L. (1950): Speciation in ancient lakes, *Quert Rev. Biol.* 25-30 – 60, 131-176.

Castenholz, R.W. (1969a): Thermophilic Cyanophytes of Iceland and the Upper temperature limit, J. Phycol. **5:** 350-358.

Castenholz, R.W. (1969b): Laboratory culture of thermophilic cyamophytes. Schweiz Z., Hydrol. **32:** 538-551.

Choudhary, L.K. (1996 : Eco-biology of fishes and the status of fisheries of the river Kosi, Ph.D. Thesis. Bhagalpur University, Bhagalpur, India.

Danchakoff, V. (1916), Anat. Ext. **10**: 397.

Das, K.N. (1968): Soil erosion and the problem of silting in the Kosi catchment. J. Wat. Cons.India. **!6(3-4)** : 62-64.

Datta Munshi, J.S., J. Datta Munshi, L.K. Choudhary and P.K. Thakur (1991): Physiography of the Kosi River basin and formation of wetlands in South Bihar: a unique freshwater system. J. Freshwater Biol. **3(2):** 105-122.

Datta Munshi, J.S. and M.P. Srivastava (1988): Natural history of fishes and systamatics of freshwater fishes of India, Narendra Publishing House, Delhi, 403 pp.

Datta Munshi, J.S. and S. Choudhary (1996): Ecology of *Heteropneustes fossilis* (Bloch), an air-breathing catfish of South East Asia. Narendra Publishing House, Delhi 174 pp.

Datta Munshi, J.S. and Datta Munshi, J. (1985): Nature of freshwater resources of Bihar. In Recent Acvances in Zoology (C.B.L. Srivastava and S.C. Goel, eds) pp. 127 – 139, Publ. Rastogi Company, Meerut.

Dehadrai, P.V. (1962): Progress of the work on the techniques of culture of air-breathing fishes in swamps in Bihar, Assam and Mysore comes of the project. Second workshop field at Patna. Dec. 20-21, 1-26 pp.

Dehadrai, P.V. and Tripathi, S.D. (1976): Environment and ecology of freshwater air-breathing teleosts. In *Respiration of amphibious vertebrates* (G.M. Hughes, ed) pp 39-72, Academic press, London and New York.

Doan, C.A., J. (1932): Lao and Clin. Med, **17,** 837.

Downey, H. (1938): Handbook Hematology, New York.

Drouet, F. (1938): Myxophyceae of the Yale North India Expedition, collected by G.E. Hutchinson, Trans Amer Micros, Soc **57:** 127-131.

Fodged, N.(1948): Diatoms in water courses in Funen IV-VI. *Dansk, Bot.Ark.***12,** 1-110.

Ganapati, S.V. (1959): Ecology of tropical waters, Proc. Symp. On Algology. New Delhi (ICAR) 200-218.

Gonzalves, E.A. (1947): The algal flora of the hot springs of Vajreswari near Mumbai. J. Univ. Bombay, **1€:** 22-27.

Goddman, C.R, Masson, D.T. and Hobbie, J.E (1967): Two Abtarotic desert lakes, *Limnol.Oceanogr.* **12:** 295-311.

Gupta, R.S. (1972): Blue-green algal flora of Rajasthan. *Nova Hedwigia.*

Hamre, C.J. (1947): J. Lab and Clin. Med., **32:** 756.

*Hendelberg., J. (1961): The freshwater pearl mussel – Report of the Institute of Freshwater Research Drottninggholm,* **41:** *149-171.*

Healy, J.M. and Lester, R.J.G. (1991): Sperm ultrastructure in the Australian oyster *Saccostrea commercialis* (Iredale and Roughly) *Bivalvia Ostreoidae* – J. Moll. Stud., **57:** 219-224.

Hora, H.L. and Pillay, T.V.R. (1962): Hand book of fish culture in the Indopacific region *FAO Fish Biol. Tech pap.* **14:** 204 pp.

Hussainy, S.U.(1967): Studies on the Limnology and Primary production of a tropical lake, *Hydrobiologia.* **30:** 335-352.

Hustedt, F. (1939): *Diatomeeus aus den Pyrenaen.* Ber.dtsch, Bot.Gas, **56**, 543-572.

Hutchinson, A.H.S.C Lucas and M. McPhail (1929): Seasonal variations in the chemical and physical properties of the waters of the *Strait of Georgia* in relation to phytoplankton. *Trans. Roy, Soc. Canada* 3, 177-183.

Hutchinson, G.E. (1957): A *treatise on Limnology Vol.I,* Geography, Physics and Chemistry, John Willey and Sons, Inc. New York. 1015 pp.

Jhingran, V.G. (1979): *Fish and Fisheries of India.* Hindusthan Publishing Corporation, Delhi, India. 958 pp.

Laal, A.K. (1981): Studies on the ecology and productivity of swamps in North Bihar in relation to production of fishes and other agricultural commodities, Ph.D. thesis, Bhagalpur University 163 pp.

Maximow, A.A. and Bloom, W. (1960): A Test Book of Histology, VII Edition, W.B. Sundars Co., Philadelphia and London.

Maximow, A. (1927): Bindegewebe and blutbidende Gewche, Handb. D. mikr., Anat., Part I (V. Mollendorff) Berlin, 1927.

Morton, B. (1982): Some aspects of the population structure and sexual strategy of *Corbleula* of *fluminialis* (Bivalvia: Corbiculacea) from the pearl River, People's Republic of China – J. Moll. Stud., **48:** 1-23.

Morton, B (1986): The population dynamics and life history tacties of *Pisidium clarkeamum* and *P. annandalet* (Bivalvia: Pisidiidae) sympatric in Hong Kong – J. Zool. Lond. (A), **10:** 427-449.

Munshi, J.S.D. (1960): Ind. J. Zoot., **1(3),** 135 + pls, I-XVII.

Nagabhusanam, R. and Bidarkar, D.S. (1990): Reproductive biology of Indian rock oyster *Crassostrea culcullata* – India J. Fish, **24 (1&2):** 135-142.

Naiman, R.J. (1976): Primary production, standing stock and export of organic matter in a Mohave Desert Thermal stream. Limnol. Oceamogr. **21:** 60-73.

Nair, G.U. (1964): On some *Nostocaceae* of Kanpur. Proc, Nat.Acad.Sci.India, 34:232-236.

Nasar, S.A.K. and Datta Munshi, J. (1974): Seasonal variations in the physic-chemical and biological properties of a tropical shallow pond, Jap. J. Ecol. **24(4):** 255-259.

Nasar, S.A.K. and Datta Munshi, J. (1972): Studies on the macrophytic biomass production and fish population in an abandoned pond of Bhagalpur, Bihar. Bh. U. Nat. Sc. Jour **4(1):** 8-16.

Nasar, S.A.K. (1975): Studies on some aspects of pond ecosystem at Bhagalpur. Ph.D. Thesis, Bhagalpur University, Bhagalpur.

Odum, E.P. (1957): The ecosystem approach in the teaching of ecology illustrated with sample class data. *Ecology,* **38:** 531-535.

Odum, H.T. (1957): Trophic structure and productivity of Silver Springs, Florida, Ecol, Mongr, **27:** 35-112.

Patil, V.Y. and Bal, D.V. (1967): Seasonal gonadal in the adult Freshwater mussel, *Parreysia favidens* var. *Marcens (benson)* – J. Shivaji Univ. (Sci.), **65 (1):** 26-33.

Patil, V.Y. and Bal, D.V. (1976): Seasonal changes in chemical composition of the freshwater musse. *Parreysia favidens* var. *Marcens (benson)* – J. Shivaji Univ. (Sci.), **16:** 59-63.

Patra, A.K., Biswasm N., Ojha, J. and Munshi, J.S.D. (1978): Circadian rhythum in bimodal oxygen uptake in an obligatory airbreathing swamp *Amphipnous* (*Monopterus cuchia* (Ham), Indian, J. Exp. Biol. **16 (7):** 808-809.

Philipose, M.T. (1959):Freshwater phytoplankton of Inland Fisheries. Proc. Sympon.Algology, 243-271.

Prasad, B.N. and Srivastava,P.N.(1965): Thermal algae from Himalayan hot spring Proc.Nat. Inst. Sci Indian. **31B:** 4-53.

Prescott, G.W. (1951): Algae of the Western Great Lakes area. Cranbrook Inst. Sci. Bull 31:946.

Rai, D.N. and J.S. Datta Munshi, (1979): Observations on diurnal changes of some physico-chemical factors of three swamps of Darbhanga. Comp. Physicol. Ecol. **4(3):** 52-55.

Rai, D.N. and J. Datta Munshi, (1979): The influence of thick floating vegetation (water hyacinth *Eichhornia crassiepes)* on the physico-chemical environment of a freshwater wetland Hydrobiologia, **62:** 65-69.

Rai, D.N., P.K. Verma and J. Datta Munshi (1980): Interactions between floating *Trapa bipinosa* and submerged vegetation community in a fish pond of Bhagalpur, Polskie archiwur Hydrobiologia, **27:** 137-142.

Rai, D.N. and J.S. Datta Munshi (1982): Ecological characteristics of Chaurs of North Bihar. In Wetlands II, Ecology and Management. Gopal, B. Turner, R.E.Wetzal, R.G. and Whigham, D.F. (eds) Int. Sci. Publ.and Nat.Instt. Ecol., Jaipur, India, 89-95 pp.

Rai, D.N (1980): Ecological structure of certain swamps of Darbhanga (Bihar), Ph.D. Thesis, Bhagalpur University, India.

Rai, D.N. and Datta Munshi, J. (1979): The influence of thick floating vegetation (water hyacinth) (*Iichhornia crassipes*) on, the physico-chemical environment of a freshwater wetland. Hydrobiol. **62 (1)**: 65-69.

Rai, D.N. (1980): Ecological structure of certain swamps of Darbhanga (Bihar), Ph.D. Thesis. pp. 164 Bhagalpur University.

Reid, G.K. (1961): Ecology of Inland waters and Estuaries. *V*an Nostrand Reinhold Publishing Co., New York, 375 pp.

Reid, G.K. and Wood, R.D. (1976): Ecology of Inland waters and Estuaries, ed, seed (New York, London, Toronto D. Van Nostrand Co.) 485 pp.

Rodhe, W. (1969): Crystallization of eutrophication concepts in Northern Europe. In Eur(o)phication, causes, consequences and correctives pp. 50-64. National Academy of Science. Washington, D.C.

Ruttner, F. (1931): Hydrographic and hydrochemische Beobactungen and Java, Sumatra und Bali, *Arch. Hydrobiol*. Suppl. 8: 197-454.

Ruttner, F. (1953): Fundamentals of Limnology (Translated by D.G. Frey and F.E.J. Fry) University of Toronto, 242pp.

Saha, S.K., Datta Munshi, J., Munshi, J.S.D. and Sarkar, H.L.(1978): Limnobiotic survey of thermal Springs of Bhimbandh. *Geobios*. 5: 205-207..

Saha, S.K. and Datta Munshi, J. (1982): Primary production and standing crop and efficiency of net production in thermal Stream, trop. Ecol. **23(2):** 300-339.

Saha, S.K. and Datta Munshi, J.,(1983): Algal flora of Sitakund and Bhimbandh thermal Springs of Bihar. Boil. Bull India **5(3):** 257-261.

Schwmmer, M. and Schwmmer, D. (1955): The role of Algae and plankton in medicines Grun and Stratton, New York, 85pp.

Sharma, V.P., S.R. Roy and D.N. Rai (1983): Aquatic mollusks on Bhagalpur., Biol. Bull. India **9(2)** 147-155.

Sharma, V.P.,S.P. Roy and D.N. Rai (1983): Aquatic mollusks and Bhagalpur, Biol.Bull. India, **5(2);** 147-155.

Singh, A.K. (1986): Characteristics of silt depositionin the Kosi river basin, Nat. Geo **21(1)**: 107-115.

Smith, D.G. (1979): Marsupial anatomy of the demibranch of *Margaritifera margaritifera* (Lin.) in northeastern North America (Pelecypoda: Unionacea), - J. Moll. Stud., **45:** 39-44.

Sonnereat, P. (1774): Observations of unphenomena singuler surds poissan qui vivent dams uneeanquia soix-ancient degree chaleur. Observations surla physique. Jour. De Phys. **3:** 256-257.

Sreenivasan, A. (1964): The limnology, primary production and fish population in a tropical pond.*Limnol. and Oceangr*, **9:** 391-396.

Stockner, J.G. (1967):Observations of thermophilic algal communities in Mount Rainer and Yellowstone National Parks. Limnol. Oceamogr. **12**: 13-17.

Thienemann, K. (1925): *Die Binnengewasser Mitteleuropas*; Eine Limnologische Einfuhrung Die Binnengewasser **1:** 1-225.

Verma, P.K. and Datta Munshi, J. (1987): Plankton community structure of Badua Reservoir, Bhagalpur (Bihar), Trop. Ecol. **28**: 200-207.

Verma, P.K. and Datta Munshi, J. (1983): Limnology of Badua Reservoir of Bhagalpur Bihar, Proc. Indian nats, Sci. Acad B. **49 (6)**: 598-609 106, 53.

Welch, P.S. (1952): *Limnology*, 2nd Ed. Mc Graw-Hill Book Co., New York, p. 538.

Woodson, B.R. (1960) A study of *chlorophyta* of James River Basin, Verginia – II. Ecol. Verginia, J., Sci. **11**: 22-36.

# Author Index

## A

Agrawal, H.P., 41
Ahmed, M.R., 17, 87
Atkins, W.R.G., 93

## B

Bal, D.V., 43
Bauer, G., 41
Bidarkar, D.S., 41
Bloom, W., 3
Bloomer, H., 34, 40
Brock, T.D., 6, 44, 45, 72

## C

Castenholz, R.W. 6, 45
Choudhary, L.K, 29
Choudhary, S., 17

## D

Danchakoff, V., 3
Das, K.N, 17
Datta Munshi, J., 17
Dehadrai, P.V, 134, 138
Dehadrai, P.V, 19, 31, 95

Downey, H., 3
Drouet, F., 44

## G

Ganapati, S.V., 88, 93, 102
Gonzalves, E.A, 44
Gupta, R.S., 87

## H

Hariss, J.P, 95
Healy, J.M., 49
Hora, H.L., 95
Hutchinson, A.H.S.C. Lucas, 63, 66, 67, 93

## J

Jhingram, V.G., 137, 138
Jhingram, V.G., 17, 22, 29

## L

Laal, A.K, 19, 31

## M

Maximoh, A.A., 2
Munshi, J.S.D., 6, 17, 22, 24, 27, 29, 40, 69, 74, 85, 99, 101, 130

## N

Naiman, R.J., 6
Nassar, S.A.K, 87, 102

## O

Odum, H.T., 68

## P

Patil, V.Y., 33, 41
Patil, V.Y., 33, 41
Philipose, M.T., 87
Pillay, T.V.R., 95
Prasad, B, N, 6

## R

Rai, D.N, 24, 26, 97
Ramaswamy, L.S., 135
Rodhe, W., 71
Ruttner, F., 69

## S

Saha, S.K., 6, 102
Sharma, V.P., 29
Singh, A.K., 17, 87, 102, 130
Smith, D.G, 34
Sonnereat, P., 6
Srivastava, P.N., 6, 17
Stockner, J.G., 6
Sundarraj, B.I., 135

## T

Thakur, P.K., 134

## V

Verma, P.K., 69, 74

## W

Welch, P.S., 95

# Subject Index

## A

A/I = Assimilation efficiency 118
Adductor muscle 38
Aerial respiration 2
Air-breathing fishes 1, 26
Amphibious 85
A*nabaena* bloom 86
Anadomers eygenea in respect of sex and gonadal changes, 33
Annelida, Molluscs 6
Annelids 100
Aquatic ecosystem 42
Aquatic macrophytes 85
Arctic, Antarctic and alpine lakes 111
Arsenic 43
Australia of freshwater 73

## B

Bacillariophyceae 56
Bacillariophyceae 105
Bacterial denitrification 98
*Bellamya* 120

Benthos. 28
Bhangchu 16
Bhimbandh 46, 51, 52, 53
Bicarbonate alkalinity 8
Biology of thermophilic algae 45
Biomass 120
Biomass estimation of algal mat 45
Biotic component 98
Biotic components 28
Blooms of *Oedogonium* 86
Brahmkund
Byssus thread 40

## C

Cage culture 135
Carbonate alkalinity 8
Carbonate alkalinity 86
Carnivorous fishes 118
*Cedesmus quadricaudata* 91
*Ceratophyllum demersum* 89
Channel (Champanala 111
Chaurs 96, 97

Chemolithotrophic Bacteria 124

Chiblung 16

*Chirononids* 86

Chloride 102

Chloride concentration 105

Chlorophyceae 105

Chlorophyceae 57

Chlorophyceae 6

*Chlorophyceae – Volvox auereus;* 91

Cirrhinus mrigala 79

Cladocerans 107

*Closterium acerosum* 91

*Conferva thermalis* Birdwoodi 44

Cortical region 38

Cyanophyceae (Blue green algae 27

**D**

Decomposition of pyrite in the generation of H2S 49

Detritus chain 133

Dilute sodium chloride-bicarbonate solutions 43

Doped" or "polluted 95

Duck breeding ponds 55

Dzakarchu 16

**E**

Ecodevelopmental strategies 31

Ecological Pyramids 130

Endothelium 5

Equilibrium carbon-dioxide 98

Europhic lakes 69

Eutrophication of wetland/swamps 19

**F**

Fauna 58, 85

Fecundity 34

Flora 85

Flora of thermal springs 50

Fluoride 43

Full overturn 69

**G**

G/A =Net growth efficiency, K2 118

G/I = Gross growth efficiency, K1 118

Gasainthan 15

GC/m2/day 86

Glaciers 85

Glochidium larvae 41

Gonads of Indomala caerulae (Lea) 40

Green algae 26

**H**

Hathkatora Pond 88

Hemopoiesis 3

Herbivorous 112

High concentrations of humic acid 72

Honey-combed appearance 5

Hwang-Ho of China 17

*Hydrilla verticillata* 26

*Hydrilla verticillata, Potamogeton crispus,* 99

*H*ydrobiological 102

Hypertrophy and hypenpla 3

Hypo-and hypercarbic conditions 100

Hypophysation technique 135

**I**

Immemorial the inhabitants of the country 52

Induced Breeding 135

Intravascularly 5

Isothermal condition 76

**K**

*Kalikosi* 17

*K*ausik of the legends 15

Kusheswarsthan 14

# L

*Labeo rohita* 4, 79
*Lamellidens corrianus* 18
Light microscope 34
Lithosphere by synthesis from oxygen and hydrogen 47
Littoral plants 71
Littoral zone 84
Lumen of the blood vessel 5

# M

Macerated tissue 54
Macrofauna 88
Macroinvertebrates 26, 28
Macrophytes 26
Macrovegetation 88
Macrovegetation in water 96, 98
Magmatic fluid 48
Makhana 19
Makhana' (*Euryale ferox* 96
*M*aqcrophytes 87
Maradhars' (Dieying Channels 32
Mayflies (*Ephemeroptera*), *Caddisflies* (Trichoptera 86
Menchu 16
Mercury Celsius thermometer 104
*Meromictic* 70
*M*esenchymal cells. 3
Meteoric water 47
Microbial ecology 42
Midges (*Diptera*) 86
Mollusk- 18
Mosquito larvae 86
Mosquito larvae 86
Murrels 100
Myeloid organs 3
Myxophyceae 105
Myxophyceae, 6

# N

*Najas graminea* 26
*Narcens* 18
Nematoda Insects 6
Nitrate-nitrogen of surface 102
Nitrogen fixation by benthic autotrophs 31

# O

Oligotrophic lakes 71
Ova 35
Ox-Bow 84
Oxidation 85
Oxygen uptake 1
Oyster 33

# P

Paddy-cum-fish culture 31
*P*arreysia favidens 18, 33, 34
Peninsula since 48
Periphyton 26, 28
Perivitelline space 38
PH value 77
Phenolphthalein and methyl 104
*P*hormidium sp. 8
Phosphate phosphorus 102
Photosynthesis 1, 87, 92, 109
Phungchu 16
Phytoflagellates 86
Phytoplankton 26, 78
Phytoplankton, zooplankton, periphyton 98, 99
*P*ila globosa 120
*P*inax 18
*P*olysaccharides. 123
*P*ondorina morum; Pediastrum simplex 91
*P*ostreproductive period, 41
Postulated reproductive strategy 40
*P*otamogeton crispus 26

*Potamogeton crispus, Nejas graminea.* 99

*Potomageton* 113

Prosobranchs 120

### R

Radio iosotopic tracer 110

Radioactive elements 49

Radioactive properties 44

Recent Volcanism 48

Reproductive cycle 41

Riverine-fisheries 29

Rotifera 28

### S

Saptadhar 7

Seasonal stream 'Badua' 74

Sediments (autochthonous) 98

*Senescent swamps* 96

*Se*wage 85

Shallow pond 94

Shichu 16

Silicate 102

Silicate concentration 105

Singhara (*Trapa bispinosa*) 19

Singhara' (*Trapa bispinosa, T. natus* 96

*S*kandha Puran 15

Spawner 135

Spectroscopical analyses 49

*S*pirogyra 86

*S*tabilization 85

Stem cells 38

Submerged vegetation 24

Sustainable hydrological cycle 20

Swampy wetlands 22

### T

Tatalpani or Taptapani 53, 54

Thermal gradient 7

Thermal spring 46

Thermocline 69

Thermocline Zone 67

*T*hiothrix and *Thiobacteria (Thiobacillus)*: 124

Torrential rains 51

### V

Venus marcensria 33

Vitelline membrane 38

Volcanic activity 49

Volcanically active regions 47

### W

Warm *monomictic* 70

*W*ater hyacinth 97

Wetlands (chaurs) 14

Winter kill 69

### Y

Yolk particles 38

### Z

Zooplankton 26, 28, 81

www.ingramcontent.com/pod-product-compliance
Lightning Source LLC
Chambersburg PA
CBHW021434180326
41458CB00001B/265